25.00

Buildability

Buildability: successful construction from concept to completion

Trevor M. Holroyd

Published by Thomas Telford Publishing, Thomas Telford Ltd, 1 Heron Quay, London, E14 4JD.

www.thomastelford.com

Distributors for Thomas Telford books are
USA: ASCE Press, 1801 Alexander Bell Drive, Reston, VA 20191-4400, USA
Japan: Maruzen Co. Ltd, Book Department, 3-10 Nihonbashi 2-chome, Chuo-ku, Tokyo 103
Australia: DA Books and Journals, 648 Whitehorse Road, Mitcham 3132, Victoria

First published 2003

A catalogue record for this book is available from the British Library

ISBN: 0 7277 3207 2

Throughout the book the personal pronouns 'he', 'his', etc. are used when referring to 'the client', 'the designer', etc. for reasons of readability. Clearly, it is quite possible these hypothetical characters may be female in 'real-life' situations, so readers should consider these pronouns to be grammatically neuter in gender, rather than masculine.

Typeset by Alex Lazarou, Surbiton, Surrey
Printed and bound in Great Britain by MPG Books, Bodmin, Cornwall

Dedication

This is for the many experienced engineers and foremen who perhaps aged prematurely as they struggled to teach engineering commonsense and practicality to an older generation. Hopefully we did learn.

Now, in an age where time is at a premium, it is not easy to gain the experience which we accepted as the norm. This contribution might help.

Foreword

It is a great pleasure to be asked to write the foreword to *Buildability: successful construction from concept to completion* by Trevor Holroyd. I knew Trevor at the beginning of his career when he was starting to develop the skills needed to make a good construction manager. His wide knowledge and understanding gained from experience are clearly evident from his book. The construction industry has been making many changes to improve its efficiency. These changes aim to reduce the capital cost and to shorten the overall time for construction, while absorbing an increase in standards, and a better working environment. One of the ways to increase efficiency is by getting better at buildability.

There are many people who participate in a construction project. A client, public or private, starts the process and establishes a basis for a building or project, studies are made to examine various options and sources of funds are examined, until the project is proved viable. Then engineers and architects are appointed who are responsible for the design and specification, with quantity surveyors acting to produce the tender documents. After the tenders are received and adjudicated, the successful contractor is appointed. Work is undertaken by the contractor's workforce and by sub-contractors; plant and materials are supplied by manufacturers. An experienced management team will be appointed to supervise and control the works.

There are many hands and minds which contribute to the project and they all need direction and coordination. In the design phase, much useful thought can be given to buildability by considering methods of construction, safe access, the location of temporary works and the phasing of the works. Careful planning of work in the construction phase is most important, and much benefit is gained from an experienced and well-trained workforce. A craft-trained foreman can contribute considerably to improving buildability.

The process of construction is complex, with multiple operations taking place simultaneously. The design planning and coordination in the detail of the work gives a good opportunity to develop better buildability. There are many excellent examples in this book, which will be of help to readers. It fills an important gap in construction literature and makes a positive contribution to improve buildability and hence efficiency. I am sure it will benefit its readers and the construction industry generally.

Sir Frank Gibb

Preface

For a number of years the construction industry has devoted large amounts of time and effort on the study and understanding of the various conditions of contract the industry uses.

At the same time we seem to have devoted relatively little time to the learning of the practical skills which are needed to carry out the work. We now lack adequate numbers of craftsmen and experienced supervisors and there is a question to be asked of the adequacy of the practical skills of our engineering staff.

Yet the types of contract which go well and leave a satisfied client and contractor tend to be those which are:

- properly planned, programmed and designed
- buildable and easy to construct in a safe manner.

The conditions of contract for such jobs require little scrutiny as mutual success is assured. For success we might consider:

- the reports by Sir Michael Latham (1994) and Sir John Egan (1998) which give guidance on how improvements can be made, particularly insofar as team working and new techniques are concerned
- the Construction (Design and Management) Regulations 1994 and other regulations, properly applied, can bring substantial improvement, not just to safety, but to the construction process itself
- practical skills, knowing how to carry out tasks the easy way as opposed to the hard way — this needs to be included in a project from the onset. If it is not, then costs will tend to increase, safety erode, programmes extend and we will have problems.

The last point is why this book has been written. It is not intended to be a volume of high learning, or to cover every construction activity like an encyclopaedia. Rather, it is a list of simple points that have been put forward by a variety of experienced people in the belief that, if they are implemented, the construction work will be easier.

If we could use feedback throughout the supply chain, noting the problems of those at the point of work and then trying to eliminate those problems in future, then we could make real progress.

Author's note

The information in this book draws on my experience and the experience of colleagues that I have worked with over the years, and this is what is inferred by the use of 'we' and 'our' in the text.

Trevor M. Holroyd

Contents

Chapter 1

The state of the industry

The construction industry is a major employer of people and a key element in the gross domestic product (GDP) of the UK. The employed total some 1·4 million and the industry turnover of 2000 is some £60 billion or about 10% of the GDP. As the provider of the national infrastructure fabric — the buildings we work in — and homes we live in, it occupies a key position in the national economy.

A vibrant economy would logically support a highly modernised industry, staffed by well-qualified and trained people who are equipped with the best possible tools to do the job. The industry should turn out quality products at reasonable cost and on time. The workforce should be well paid and highly motivated. The engineers and managers who control the industry should be respected and reasonably rewarded.

In fact, the reverse of this is perhaps nearer the truth. Some areas of the industry do produce high-quality work on time and some managers and workers are well paid. There is, however, a suggestion that the industry could and should be more efficient, the workforce should be better trained, management more effective and the final product — the infrastructure around us — should be of a higher quality, more durable and represent better value for money.

These are sweeping statements to make. While we may not fully agree with the words used, few would disagree with the statement that the industry has a poor image as a whole, and the many people who work *extremely* hard within it will often feel that their efforts are poorly rewarded financially and that they receive little professional respect from the nation as a whole.

Successive governments have been concerned by this performance and major reports have been commissioned to study the problem. These were, notably, the Lathan Report (1994) and the Egan Report (1998).

Constructing the team — The Latham Report

Constructing the team by Sir Michael Latham was published in July 1994. It is a governmental and industry review of the methods of procurement and the contractual arrangements, which apply in the UK construction industry. The report principally concerns itself with the performance of the industry, increasing efficiency, operating on a basis of fairness to those contributing to construction work, and encouraging teamwork to get best results.

The report identifies clients as the driving force for the implementation of its findings. It states that clients have a responsibility to commission projects of which we could be proud. They should promote good design and consider carefully energy and maintenance costs as well as appearance and effective use of space in a building. The lifelong costs of a project require consideration.

The industry does, however, rely on the wider economy to prosper. If the flow of work into the industry is less than its capacity, the consequences will be that:

- companies will reduce staffing levels, and some will close down
- consultant's fees will be reduced and the final design outcome may suffer — consultants will find it difficult to progress the business
- contractors' prices will be too low to be economical — payments by contractors will be delayed, training and employment will also suffer, and quality and best practice performance on site is likely to be affected
- training and education within the industry will suffer
- there will be little money available for research and development or for enhancing the public image of the industry.

Anyone involved in the industry in the recent past will recognise these symptoms.

Contract strategy

Clients commissioning projects will normally look for factors such as:

- value for money
- good to look at
- free from defects

- delivered on time
- fit for the intended purpose
- proper guarantees apply
- running costs are reasonable
- the project is reasonably durable.

Anyone going into a shop to buy a car, or a television set or a washing machine, will expect goods which satisfy these criteria. Yet in construction we are all too often perceived as failing to deliver and this gives rise to client criticism.

Clients are often uncertain of what they want and therefore give an inadequate brief to their consultant/contractor. The client then clarifies his needs, and makes changes to the work. This leads to increased costs, programme delays and a dissatisfied client.

Latham suggests that once a client identifies the need for a project he should first devise a contract strategy based on the amount of risk he is willing to take. If clients are prepared to take all the risk, then a management type of contract will be beneficial. At the other extreme, a package deal will see the contractor taking the risk. Once the strategy is decided by the client, then he should approach a designer or a contractor or other party. Who he approaches will be determined by the strategy adopted.

Once the contract strategy is agreed and the client wishes to proceed, the project brief, which covers the basic objectives of the client, and the design brief, which covers the specific requirement, can be prepared.

Clients who know what they want, who do not change their minds, and who state clearly what they want, tend to procure projects that do represent value for money. On the other hand, clients who are uncertain and change their minds tend to pay more for a less satisfactory outcome. Contractors tend to perform better on the former and worse on the latter case.

The design process

If a project is to be successful it is essential that it is managed effectively. The report suggests the following.

(*a*) A lead manager is appointed for the design process and should head an integrated design team. It can be very beneficial to appoint a single project manager to deal with the contractors and the various consultants employed on the project. The project manager acts as the Client's representative. The *New Engineering Contract* (NEC) (ICE, 1993) recognises this.

Some clients have their own in-house project teams. Many will employ outside agents. A client may have good experiences of such appointments and have particular preferences.

It is important that the proposed structure performs as the client wants it to and that it meets its objectives. The design team leader must ensure that the client understands the proposals fully.

(*b*) Consultants need to be coordinated. Their appointment documents should form an interlocking matrix and have a clear relationship with the contract.

Total design/construct cover is needed for the project with no gaps between the responsibilities of those concerned. By ensuring that an interlock occurs between appointments this can be achieved insofar as the contract is concerned.

(*c*) A detailed checklist of the design requirements should be included in the appointment documents of consultants. This should also be set out in the main contract documentation.

Everyone then knows what is required. The checklist can be scrutinised, changed or discussed as necessary. The greatest advantage is probably the agreement by all concerned that the list is complete and fully correct — that everything is covered before the design process starts in earnest.

An incomplete list, or a list that is provided at a later stage, can have consequences which will most likely be negative.

(*d*) Particular care should be taken over the integration of building services and the avoidance of 'fuzzy edges' between consultants and specialist engineering contractors. The latter would include all specialist services to a structure — heating, ventilating, air conditioning, for example — as well as specialist sub-contractors who may have a design responsibility — piling, suspended floors and ceilings, curtain walling, for example.

Too many managers in the construction industry have a less than adequate knowledge of building services. Projects run into difficulty because of poor coordination, sometimes caused by this lack of knowledge.

'Fuzzy edges' occur when the interfaces between items provided by different specialist contractors are not correctly defined. Elements clash in terms of position, or are incompatible, and variations are introduced. Considerable expense tends to be incurred when this happens.

Ensuring that the various design elements are correctly integrated in respect of each other is common sense.

(*e*) Project information must be coordinated. We need to ensure that there are no gaps in the construction information provided.

The client may need to commence construction before the full design is available, or to change his requirements during the construction process. The system adopted must allow for this. Best practice is, nonetheless, that all projects are properly pre-planned.

Computer software is available which enables the proposals of the various contractors to be superimposed on each other, often as a virtual three-dimensional layout. This makes it easier to prevent a clash between the various provisions.

(*f*) The different stages of the design should be 'signed off' when they have been achieved.

The signing off would logically be by a senior experienced person. A predetermined signing off process allows regular, comprehensive checks to be made in a knowledgeable manner. The intention is to eliminate errors. This is likely to reduce the costs associated with the elimination of errors.

The contract choice for clients

The report suggests that the most effective form of contract in modern conditions should include the following.

1. A specific duty for all parties to deal fairly with each other, and with their sub-contractors, specialists and suppliers, in an atmosphere of mutual cooperation.
2. Firm duties of teamwork, with shared financial motivation to pursue those objectives. These should involve a general presumption to achieve 'win-win' solutions to problems which may arise during the course of the project.
3. A wholly interrelated package of documents that clearly defines the roles and duties of all involved, and that is suitable for all types of project and for any procurement route.
4. Easily comprehensible language and with guidance notes attached.
5. Separation of the roles of contract administrator, project or lead manager and adjudicator. The project or lead manager should be clearly defined as the client's representative.

6. A choice of allocation of risks, to be decided as appropriate for each project but then allocated to the party best able to manage, estimate and carry the risk.
7. Taking all reasonable steps to avoid changes to pre-planned works information. Where variations do occur, they should be priced in advance, with provision for independent adjudication if agreement cannot be reached.
8. Express provision for assessing interim payments by methods other than monthly valuation, i.e. milestones, activity schedules or payment schedules. Such arrangements must also be reflected in the related sub-contract documentation. The eventual aim should be to phase out the traditional system of monthly measurement or remeasurement but meanwhile provision should still be made for it.
9. Clearly setting out the period within which interim payments must be made to all participants in the process, failing which they will have an automatic right to compensation, involving payment of interest at a sufficiently heavy rate to deter slow payment.
10. Providing for secure trust fund routes of payment.
11. While taking all possible steps to avoid conflict on site, providing for speedy dispute resolution if any conflict arises, by a predetermined impartial adjudicator/referee/expert.
12. Providing for incentives for exceptional performance.
13. Making provision where appropriate for advance mobilisation payments (if necessary, bonded) to contractors and sub-contractors, including in respect of off-site prefabricated materials provided by part of the construction team.

The NEC is stated to cover virtually all these assumptions of best practice.

The report recommends the making of proper payments to sub-contractors, ending the procedure of 'pay when paid'. The Housing Grants, Construction and Regeneration Act 1996 Section 113 implemented the necessary actions to control the situation of 'pay when paid'.

Selection and tendering procedures

The report identifies five issues.

1. Professional consultants should be selected on a basis that properly recognises quality as well as price.

Money rules the commercial world we live in. All businesses must tailor their product to fit the price customers are prepared to pay. It is not sensible to expect consultants to be able to give the best possible service for the lowest price. Some consultants also have better skills in some areas than others.

The sensible approach is to seek a quality consultant, one with the appropriate skills and adequate available resources to do the work required, perhaps a consultant already proven in the eyes of the client, then to negotiate a sensible fee for the required work.

2. The need for a lead manager.

   This is reinforcement for the lead manager needed to cover the design process which was noted earlier. Someone should always be available who has the knowledge, capacity and empowerment to ensure that all things are considered in seeking the best deal.

3. Contractors should not be required to undergo excessive or burdensome qualification procedures.

   Any contractor can put together a bulky, and impressive, qualification document. The document itself may prove nothing.

   Some contractors are good on big contracts, some on small ones. Some contractors will be busy and have their resources committed, others will be actively seeking work.

   Seek the right contractor, one who can and will provide a quality job, and minimise the paperwork involved. This will save time and money for preferred contractors and avoid misleading contractors who are less likely to be chosen. The latter can then pursue other work more vigorously.

4. Tender lists — including those for design and build projects — should be of a sensible length. It is an expensive process to submit a tender and only one submission can be successful.

   Minimise the cost and amount of abortive effort. Recommendation 6.32 of the report gives guidance on tendering.

5. Value for money and future cost-in-use should play an important part of the selection process.

   It is sensible to select a supplier by taking into account their overall offer in addition to price. Generally speaking, how good they are.

   A good supplier will tend to give better value for money than one not so good. The overall cost of a project is the lifelong cost, not just the cost of the construction itself. In considering lifelong cost we need to look at the costs of any maintenance, upgrading or enhancement to meet changing needs, and the ultimate cost of demolition.

The cheapest construction price may give a structure with limited durability and little scope for modernisation at a later date. This could prove expensive. On the other hand, a more expensive construction could provide greater durability and be better designed insofar as maintenance and modification is concerned.

Those preparing tenders should be given adequate time to prepare their submission properly. Guidance on tender preparation is available from the Institution of Civil Engineers (ICE) in their *Guidance on the Preparation, Submission and Consideration of Tenders for Civil Engineering Contracts* (ICE, 1983).

The Latham Report notes the Banwell findings on partnering arrangements, where contractors had given good service to clients, there was scope for awarding them work without going through the competitive process. Negotiated work tends to be successful. A practical design leads to successful construction. The contractor makes a profit and the client is pleased with the construction and cost outcomes.

On a wider scale, partnering can be effective on large development programmes where expert teams have to be established and maintained for the programme duration. The nuclear power station programme was one such example. Some colleagues were involved for up to 20 years on this programme, starting as graduate engineers and ending the programme as directors in their organisations.

The service and utility industries use partnering arrangements where term contracts are set up with a single supplier. Local authorities tend to outsource their design and, later, their construction work. The Highways Agency and the Environment Agency, both major government procurers, are quickly minimising their supplier numbers, streamlining the procurement chain and, in effect, adopting a partnering arrangement.

Contractors could use a partnering arrangement when selecting their suppliers. Long-term relationships tend to be helpful. Trust replaces uncertainty, knowledge and capability is positive.

Issues which determine performance

Construction costs appear greater in the UK than in other industrial nations. Yet labour costs are among the lowest.

There is clearly scope for improvement. A strong feature of the report by Sir John Egan (1998), *Rethinking construction*, is the need to improve. Targets are set for such improvement and areas are noted where improvement might be achieved. These are considered later.

Companies can set themselves up in construction too easily. Accreditation, based on competence, would help. It is also sensible to have a system of accreditation for operatives. The Construction Industry Training Board (CITB) has long been raising the issue of the skills crisis in the industry. In our experience the following are true.

(*a*) Good craft training can turn young people from being totally lacking in craft skills to effective tradesmen in some 18 months to two years. Many trained craftsmen are able to successfully establish their own businesses at an early age. The training grants received are quite small but we did a lot of craft training and made a profit in doing so. This enabled us to successfully provide quality work for 'blue chip' clients and expand the business dramatically.

(*b*) Working on flood defences, the practice was to work long hours in summer so as to compensate for time likely to be lost in winter due to flood conditions. The granting of a 30% pay award to the operatives made a review of the working hours vital. As a financial necessity a five-day working week was adopted with minimal overtime. The results were that operative performance improved dramatically. Operatives were benefiting from a large increase in pay. The five-day week enabled them to enjoy their weekends at home and take a rest from work. We produced as much, or more, than we had done in seven-day weeks. It became clear that we had been working long hours, as much to provide decent take-home pay as anything else. The seven-day working week ended overnight. Everyone, including the business, benefited.

An effectively trained operative, equipped with a range of relevant tools of good quality will out perform a lower paid worker using fewer tools and the latter practice will cost money, not save it. The experience of the US construction industry would appear to support this.

The report considers professional education and whether it needs a greater degree of practical experience. While the words used vary, there are strong suggestions that:

- special training may be needed to meet the requirements of an improved understanding — comments from members of the

Royal Institute of British Architects (RIBA) indicate a need to widen the practical content of their educational courses
- financial and management skills should be included in modules of education common to all entering construction at an early point in their careers. There is concern within the ICE, and this is referred to in some of the topics of the professional review, that engineers are not sufficiently skilled in management or financial matters. They remain managers in organisations rather than leaders of them.

It must be difficult, for a young designer with little practical experience of construction, to prepare drawings which incorporate a high degree of buildability. The designer, preparing details which will be assembled by a variety of tradespeople, will require progressively more experience than a single trade specialist.

There appears to be an acknowledged lack of understanding of the service requirements of structures. Many engineers openly admit to not knowing the requirements of contracts for which they are responsible. They tend to put the drawings which detail the requirements to one side and leave it to the specialist sub-contractor.

The tendency within the industry to have more engineer-trained staff and fewer craft-trained foremen is creating supervisory problems. Too many managers and designers are unaware of the requirements of specialist and other sub-contractors. As a result, they find it difficult to manage the sub-contractors effectively on site. Designs are also less effective than they might be.

Work is, however, being carried out in some areas to rectify such deficiencies. The College of North West London has 5500 students in its Faculty of Technology. Of these, some 4000 are studying topics relating to the built environment. The large number of students is supported by a large number of courses. This enables a student to pursue a wider achievement of knowledge than would occur in a small unit. National Standards are taken as the minimum provision. The college recognises that many of those studying will later become self-employed and thus require a knowledge of communication skills (how to write letters and speak at meetings, for example). They will need to have an adequate command of the language. Students are therefore offered courses relevant to their main requirement so as to enable them to be self-sufficient. To stand on their own feet.

Staff are encouraged to attend courses that provide them with a full and up-to-date knowledge of best current practice. This is then introduced into the curriculum as quickly as possible. The challenge for people to improve themselves covers the entire faculty.

University graduates attend some courses and study alongside craftsmen seeking new skills. The clear intention is to provide students with the necessary skills to do their job properly, whatever those skills may be. For example:

- a plumber will be given a knowledge of electronics and computers — he will need all these skills in the modern occupation
- a floor tiler will be offered a knowledge of plastering and thus have wider opportunities
- an electrician may have a knowledge of soldering, or information technology (IT) or electronics
- specialist trades (leadworkers, for example) may well also have a degree of skill in tiling.

These are examples of an approach to multi-skilling which will offer wider employment opportunities to individuals and a wider skill base to the employer. Engineering is regarded as integral to construction and widens scope further still.

The provision of quality work is seen as cheap. The lack of quality, expensive. This gives an immediate motivation to provide the best.

The Faculty confidently states that the provision of the relevant education and training is profitable. Our experience led to the same conclusion.

The Latham Report identifies the serious concern felt within the industry at the low level of investment in research and development. In general terms, expenditure needs to be doubled from its current level of 0·1% of GDP.

How increased funding can be achieved remains to be answered. Existing initiatives should be monitored and further data collected before a decision is reached, the report suggests.

Teamwork on site

The review states that most of the input received from the industry was concerned with the relationship between main contractors and specialist sub-contractors. A contractor's survey of sub-contractors asked questions such as:

- are your tenders dealt with fairly?
- does the main contractor indulge in Dutch Auctions?
- how comfortable are you with the main contractor's terms and conditions?
- are you fairly treated on financial matters?
- are contra charges handled properly?
- do you feel part of the team?

Responses to such questions revealed ratings that were less than acceptable. The Housing Grants, Construction and Regeneration Act 1996 addresses some of these issues, largely by ruling that fairness *will* prevail. Recommendations resulting from the survey were as follows.

1. Better relations should be developed through partnering or partnership arrangements — this is now occurring in a major way between clients and their suppliers.
2. Involve sub-contractors earlier to achieve project objectives, and develop greater team involvement though the project lifecycle and beyond. Use their specialist knowledge to reinforce your existing 'know how' to get it 'right first time'. It is far cheaper to solve problems pre-start than during the construction phase. Speaking on this issue at a conference a delegate with experience of Japanese practice made the following comment:

    > They spend a huge amount of time planning, discussing problems and looking at the requirement from every angle. Indeed, you wonder if they are ever going to start. When they do start, however, they expect *everything* to work to plan and it often does.

3. Utilise the skill and knowledge of sub-contractors more fully and better, and recognise that sub-contractors can and want to make a greater contribution. Specialists have a greater knowledge of their specialism than the non-specialist. Their contribution is probably going to aim at doing the job more easily or better, or quicker. Everyone will benefit. In addition, those making contributions come to regard the job as their own and show a vested interest in making *their* job succeed. Those same people are less likely to show the same determination to make *your* job succeed.
4. Develop a more structured, standardised and ethical approach to the procurement and management of sub-contractors. In a competitive environment it is not realistic to employ all the required skills of a business directly. If you do, then it can become difficult to alter the

sails of the business as the economy changes. Sub-contracting, where required skills are hired on a required basis, gives the contractor the required flexibility. Costs are incurred only as necessary. They do not continue when the demand ceases as would be the case using directly employed people.

Sub-contracting has become a necessary business tool. Contractors need to sub-contract successfully. The fiercely competitive pricing of the late 1980s/early 1990s forced contractors to use the lowest priced sub-contractors regardless. Few contractors would suggest this to be 'best practice'.

It is far better to use suppliers of known value, and to use them regularly. To measure their value on overall performance and not just on price. To support those who do a good job for you.

All parties to the construction process should use standard forms of contract without amendment. The main purpose served by amended forms of contract is to give one of the parties an advantage over the other. This is where mistrust starts. Latham lists actions which are deemed unfair or invalid. These include attempts to change the terms of any payments or to deny the rights of adjudication.

You can work on a basis of trust or mistrust. Trust must help job progress. In addition, fair conditions for and payments to sub-contractors must enhance their chances of doing a better class of work than would otherwise be the case. Good relationships will have a positive effect.

Dispute resolution

The construction industries of the USA and the UK are prone to adversarial attitudes and disputes. The report recommends avoidance as the best solution. To use a contract document which emphasises teamwork and partnering, where variations are priced by agreement and a built-in adjudication process is included, will help matters. Current documentation, the NEC and the JCT (Joint Contracts Tribunal) suite of contracts (*Practice note 20* (JCT, 1988) gives guidance on the appropriate form of JCT main contract) can each be suitably amended to include this.

Alternative dispute resolution would include mediation and conciliation. For very large projects multi-tiered dispute resolution may be necessary. Arbitration is noted and the hope expressed that it will be used less often. Perhaps the key to buildability is to get it 'right first time' and avoid the dispute altogether.

In fairness to the industry, the experience of most contractors is that some contracts are more difficult than others in terms of:

- construction, quality, programme
- relationships, financial outcome.

Nevertheless, contracts:

- do get finished
- are finally resolved
- only rarely do we have to resort to arbitration or other legal process to reach final agreement with our clients.

In reviewing the procurement and contractual arrangements within UK construction, Sir Michael Latham can be seen to be focusing on the need to:

- understand the requirement
- have a fair form of contract
- provide an adequate and suitable procurement process
- encourage partnerships
- use 'best practice' and 'value engineering' as far as possible.

Buildability would then be improved in its widest sense.

Rethinking construction — The Egan Report

Sir John Egan, Chairman of the Construction Task Force, delivered the report *Rethinking construction* in July 1998.

In the Foreword to the report, Sir John said that a successful construction industry is essential to us all. That we all benefit from quality structures which are efficiently constructed. While UK construction at its best is excellent, substantial improvements can be made, and it is vital that they are made.

Experience gained from other industries that have transformed themselves in recent years and from best practice construction work is considered. It is concluded that continuous and sustained improvement is achievable if we focus on the issues which matter, to give better quality, value for money and to cut wastage, for example.

The industry is challenged to commit itself to change and to work together to create an effective modern industry.

The need to improve

The industry is too big in size, in employment terms and in its position in the economy, for it to be allowed to stagnate. There is a recognised *need to modernise*.

(*a*) Profitability is unreliable and too low to sustain healthy industry development. Investment becomes difficult to obtain. Companies are forced into an often unremitting struggle to be profitable and to survive. This is at the expense of developing the business and its people for the future.

(*b*) We invest too little in research and development (R&D) and capital expenditure is too low. R&D is 20% of what it was 20 years ago and this must restrict the development of new technologies, materials and working methods. Capital investment is 33% of what it was 20 years ago. More plant and equipment is hired in, and offices are leased. While this reduces the cash requirement, it does mean regular hire/lease payments which have still to be made when times are hard.

We found that *sensible* levels of capital investment, spent on items we used almost continuously, enabled us to get excellent returns, provided we properly maintained the items we bought.

(*c*) There is a crisis in training. The number of trainees has halved since the 1970s. The number of skilled operatives is decreasing. Too few have the technical and managerial skills to get full value from new techniques and technologies.

The industry as a whole does not have a proper career structure to develop supervisory and management grades.

One contractor has excellent, experienced supervisors running major projects. The problem is that, as these are coming to retirement, their replacements are much younger, and are comparatively lacking in experience.

Another large company, although still successful, has less turnover, less profit, fewer apprentices and trainees, than it had 15 years ago.

(*d*) Too may clients still equate price (the initial tender price) with cost (final overall cost — this would include lifelong maintenance costs). What really matters is the overall cost. This is seen as one of the greatest barriers to improvement. It tends to keep all prices down and gives no encouragement to companies who use best practice or designers who offer more durable, aesthetic contents in their structures.

Client dissatisfaction

This arises when projects are delivered which are late on completion, lacking in quality and over budget. Construction is often seen to fail to meet the needs of modern businesses who must be competitive to survive. It rarely provides best value for its clients. A survey of clients revealed that clients:

- want greater value from their buildings by achieving a clearer focus on functional business needs
- have an immediate priority to reduce capital costs and improve the quality of new buildings
- believe that a longer term and more important issue is to reduce running costs and improve the standard of existing buildings
- believe that significant value improvement and cost reduction can be gained by the integration of design and construction.

In our experience, most contractors or designers would willingly assist clients to achieve these aims, especially if it is to be done on a design and construct basis. They can then utilise their construction talents to offer best value.

This gives an opportunity from the onset to seek the best design and to drive this forward to give the best end result.

Fragmentation of the industry (163 000 companies are listed) inhibits performance:

- on the positive side fragmentation leads to flexibility in dealing with variable workloads (working within the economic cycle)
- on the negative side, sub-contracting, especially when carried out on the basis of price, has:
  - brought contractual relations to the fore
  - encouraged the mistrust of the sub-contractors for the main contractor (and vice versa)
  - prevented the building of teams and the acquiring of knowledge which is essential to efficient working.

Sub-contracting needs to be encouraged to improve, by fairer contract relationships, the adoption of partnering or other collaborative arrangements, the building of long-term relationships, and the encouragement of those who provide 'best practice'.

However, positive developments are occurring.

(*a*) Partnering arrangements are increasing in number, and many seem to be getting larger.
(*b*) Initially within buildings, but now with civil structures also, standardisation, precasting and pre-assembly are steadily increasing.
(*c*) Software of computer technology is helping designers produce better designs and members of the construction team to communicate more easily and remove barriers to construction efficiency.
(*d*) Benchmarking, the measuring of a company's standards against those of its competitors, is increasing. One business has a mission statement of 'zero defects'.
(*e*) Reducing waste, and minimising handling by arranging product delivery 'just in time', is becoming more common.

The report suggests that there is great scope for improvement in the cost, time and quality of projects.

This improvement will only be achieved when buildability is maximised across the whole construction process.

The ambition for UK construction

Task Force members were selected on the basis of their expertise as construction clients and their experience of other industries — the motor car industry, steel making, retailing, offshore engineering and other areas — where dramatic performance improvements have been made.

The radical changes that have taken place in the manufacturing and service industries have been stimulated by a series of fundamental points.

(*a*) *Committed leadership*. Management believing in and being totally committed to driving forward an agenda for improvement and communicating the required cultural and operational changes throughout the whole of the organisation.
(*b*) *A focus on the customer*. The best companies are customer driven. They provide exactly what the end customer needs, when it is

needed and at the right price. Activities which do not add value are eliminated as being wasteful. In construction, each company pursues its business in its own way and, as a whole, fails to understand this issue or its solution.

(*c*) *Integrate the process and the team around the product.* The most successful businesses do not fragment their operations — they work back from the customer's needs and focus on the product and the value it delivers to the customer. The process and the production team are then integrated to deliver value to the customer and eliminate waste in all its forms.

The Task Force sees this as fundamental to increasing efficiency and quality in construction. Eliminating waste, in time for example, can pay large dividends. An integrated team will tend to comprise the most suitable elements — better knowledge, able to learn from each other, and more confident and driven as a result.

(*d*) *A quality driven agenda.* The report states:

> Quality means not only zero defects but right first time, delivery on time and to budget, innovating for the benefit of the client and stripping out waste, whether it be in design, materials or construction on site. It also means after-sales care and quality in use. Quality means the total package — exceeding customer expectations and providing real service.

Delivering a first class product by means of a first class service without inefficiencies. Top quality manufacturers can be recognised by these criteria.

(*e*) *Commitment to people.* We ensured our commitment was successful by involving everyone in the business. The reward comes when people then come to regard the business as *theirs*. They will work harder and more positively for a business they regard as theirs than they will for a business they regard as yours.

An appraisal system ensures that people are appraised as often as necessary. The outcomes of each appraisal are:

(*i*) the agreed actions the business would take — to improve the individual's contribution to, or their rewards from, the business (more training, increased salary, for example)

(*ii*) the agreed actions of the individual to improve their contribution to the business (improving skills, eradicating defects!) — an appraisal system can quite easily cover the supplier network

(*iii*) monitoring ensures that the actions agreed by the parties do, in fact, take place.

The scope for improvement

The report sets targets for improvement. It is suggested that 30% of construction involves the remediation of defects, that labour is only 40% to 60% efficient, and that materials wastage is at least 10%. An Organisation for Economic Cooperation and Development (OECD) report (see section 2.26 of the Egan Report) suggests that UK input costs are generally one-third less of those of other developed countries. But output costs are similar or higher.

Clearly there is scope for improvement. On the basis of the figures quoted, the scope is considerable. The targets set are given in Table 1.1.

Examples of companies which have improved construction performance are as follows.

(*a*) Tesco stores have reduced the capital cost of their stores by 40% in five years. They are now targeting a further 20% reduction in costs over two years and a 50% reduction in project time.

(*b*) Argent have reduced the capital cost of office construction by 33% and total project time by 50% since 1991.

(*c*) BAA Pavement Team have reduced project time on airport runways and taxiways by more than 30%, reduced accidents by 50%, and achieved 95% predictability of cost and time in two years.

(*d*) The Whitbread Hotel Company has reduced construction time for its hotels by 40% since 1995 and costs have also been progressively reduced annually in real terms.

(*e*) Raynesway Construction Southern in a year have reduced the costs of maintaining Hampshire County Council's roads by 10%, increased turnover by 20% with the same labour force, and reduced accidents by 60%.

(*f*) The Neenan Company in Colorado have used 'lean construction' techniques over two years to reduce the time to produce a schematic design by 80% and project times and costs by 30%.

(*g*) Pacific Contracting of San Francisco have used 'lean construction' to increase their productivity and turnover as a cladding and roofing sub-contractor by 20% in 18 months.

(*h*) Neil Muller Construction of South Africa have used total quality management (TQM) techniques to achieve an 18% increase in output per employee in a year, a 65% reduction in absenteeism in four years, and a 12% saving on construction time on a major project.

Table 1.1. The scope for sustained improvement

Indicator	Improvement per year	Current performance of leading clients and construction companies
Capital cost All costs excluding land and finance	Reduce by 10%	Leading clients and their supply chains have achieved cost reductions of between 6% and 14% per year in the last five years. Many are now achieving an average of 10% or greater per year
Construction time Time from client approval to practical completion	Reduce by 10%	Leading UK clients and design and build firms in the USA are currently achieving reductions in construction time for offices, roads, stores and houses of 10–15% per year
Predictability Number of projects completed on time and within budget	Increase by 20%	Many leading clients have increased predictability by more than 20% annually in recent years, and now regularly achieve predictability rates of 95% or greater
Defects Reduction in number of defects on handover	Reduce by 20%	There is much evidence to suggest that the goal of zero defects is achievable across construction within five years. Some UK clients and US construction firms already regularly achieve zero defects on handover
Accidents Reduction in the number of reportable accidents	Reduce by 20%	Some leading clients and construction companies have recently achieved reductions in reportable accidents of 50–60% in two years or less, with consequent substantial reductions in project costs
Productivity Increase in value added per head	Increase by 10%	UK construction appears to be already achieving productivity gains of 5% a year. Some of the best UK and US projects demonstrate increases equivalent to 10–15% a year
Turnover and profits Turnover and profits of construction firms	Increase by 10%	The best construction firms are increasing turnover and profits by 10–20% a year, and are raising their profit margins as a proportion of turnover to well above the industry average

The Egan Report states:

> If the industry is not prepared to make the improvements, then we propose that the clients should take the initiative. We are already aware of the Construction Round Table's and the Construction Clients' Forum's intentions in this respect, and of the British Property Federation's customer survey. We think it is essential that any comparative data take account of user satisfaction with the buildings they occupy and with the services of the design and construction team.
>
> This then is our ambition for a modern construction industry in the UK: adoption of the model of dramatic performance improvement that other industries have followed with such success, in order to deliver the challenging targets for increased efficiency and quality that we know are achievable. In the next section we offer the industry a practical approach to doing so, through the concept of the integrated project process.

The prospective rewards are clearly worth the effort.

Improving the project process

Chapter 3 of the report asks 'What can construction learn from other industries?'. It suggests that many processes are common on all sites. While we can argue that each construction project is unique in itself, many of its elements are common to all projects — windows, doors, concrete, reinforcement, for example.

Repeated processes

If we repeat the use of the process, of concrete placing or fixing precast units and such items, then we can concentrate on improving the process. Ensuring that the concrete mix is sensibly workable makes placing much easier than a difficult to work mix. Precast work is often critically dependent on its fixing techniques. Repetitive process will improve operative skills and enable techniques to be improved as experience is gained. It is not very efficient practice to continually reinvent the wheel. Insofar as materials are concerned, we use items which we know will work, and specify them from the onset. Japanese construction makes work as repetitive as possible.

To maximise improvement we need feedback from the workplace on how the methods or materials or other items are fitting into the process. What is working well and what is not working so well. We then use the knowledge gained from the feedback to eradicate defects. To get rid of bad practice or product by change or improvement. Eradicating defects is a basic tenet of TQM and other quality based programmes.

While the greatest business improvements often arise by defect eradication it is also important to improve on your strengths. Repeating the process helps!

An integrated project process

Every project undertaken utilises a number of teams or elements and these perform better and tend to achieve better results if they work together than if they work separately, pursuing their own goal. It is the big payback of the teamwork process, to all work for each other.

The integrated project process pulls the whole team together and focuses it on a single task — delivering value to the client. By concentrating on client value, other very important issues can be resolved.

(*a*) Discussion or arguments about objectives will be minimised.
(*b*) Suggestions for project improvement will be offered or sought by all members of the team. Each suggestion employed will have a positive effect on the outcome. The total positive effect is likely to have a favourable effect on all parties involved. Less waste, improved quality, improved methods, less reworking — on designs and the work itself — will each have its benefits for the team.
(*c*) The best argument for teamworking is to ask how we would perform without it. The answer is 'not very well!'. The best team will have in its ranks the required mix of skills for the project concerned. Different elements will contribute different skills. Once we have selected the best team constituents and put them together, that team will only be given the desired result if it is focused on that result and nothing else.

Focus on the end product

The integrated process focused the team on a single task — that of delivering value to the client. Clients are identified as usually being interested only in the finished product, its cost, delivery on time, its quality and functionality — we should concentrate on these issues.

Experience within the Task Force suggests that the process can be considered as:

- product development
- project implementation
- partnering the supply chain
- production of components.

If the team can carry out a series of projects then it can progressively improve the product by learning from experience (using feedback).

Carrying out regular projects for the supermarket chains and large units in the retail sector shows that each business has its own image. Experience on one project helps on the next, teams gain required skills and keep them in place within the team for future use. Projects become easier to carry out, efficiency increases. Everyone benefits.

Product development

Product development is about continuously developing a product using people with the appropriate knowledge and a system which ensures that we get what we want using methods which are steadily improved. You need a product team that has the relevant skills to carry out the task and they need to be fully informed at all times.

Feedback is used to assess the success of an operation. The lessons learnt from the feedback are progressively incorporated into the construction activity to improve practice in all phases of the project and in the quality of the end product.

Product development comprises:

- listening to the voice of the consumer and understanding their needs and aspirations
- developing products that will exceed client expectations
- defining the attributes of a construction product and understanding how they are influenced through specific engineering systems and components
- defining projects that deliver the product in specific circumstances and setting clear targets for the project of delivery teams
- assessing completed projects and customer satisfaction systematically and objectively, and feeding the knowledge gained back into the product development process
- innovating with suppliers to improve the product without loss of reliability.

Project implementation

This is about taking the general project requirement and developing it into a specific project on a specific site. This is achieved in a manner which envisages all teams and team members working together to effectively provide a finalised project in an effective and efficient manner. A manner in which everyone contributes and gives it their best shot.

The Japanese are particularly good in this field, trying to eliminate any defect before it occurs. They regard ultimate success as making the implementation effort worthwhile.

Project implementation comprises:

- leadership of an integrated team of suppliers, constructors and designers dedicated to engineering and constructing the project
- mapping of processes, measurement of performance and continuous improvement to improve quality and eliminate waste
- development of engineering systems and selection of components to achieve product performance targets
- preplanning of manufacture, construction and commissioning
- assembly of components and sub-assemblies on site, and commissioning of the completed project
- training and development of all participants to support improvements in performance
- learning from experience and feedback into the project delivery process.

Partnering the supply chain

The supply chain role is critical to achieving sustained improvement and forward thinking. In the context of extensive sub-contracting this is perhaps more so. Partnering, correctly implemented, will provide benefits comparable to those obtained by the integrated project process. In an industry where the supply chain has traditionally been determined largely by the cheapest price submitted, partnering is a radical change.

Sensible organisations do, however, often use suppliers on a perceived best value basis. Long-term relationships between the organisation and its suppliers are seen to produce better results overall than 'first past the post, cheapest price approach'.

Our experience has demonstrated the benefits of partnering. On some contracts, where there was perhaps little choice of supplier and cost allowances were minimal, we learned the hard way. The cheapest option failed. When this occurs cost escalation can be dramatic. We quickly re-learnt the benefits of 'best value partnering'.

Partnering the supply chain comprises:

- acquisition of new suppliers through value-based sourcing
- organisation and management of the supply chain to maximise innovation, learning and efficiency
- supplier development and measurement of suppliers' performance
- managing workload to match capacity and to incentivise suppliers to improve performance
- capturing suppliers' innovations in components and systems.

Production of components

There is no reason why the right components should not be delivered at the right time, in the right order, free from defects and with minimum wastage. Management should plan to ensure that this happens.

We should be looking to improve components on an ongoing basis. Partnering/teamworking will enable this to happen more easily.

Production of components comprises:

- detailed engineering design of components and sub-assemblies
- planning, management and continuous improvement of the production process
- development of a range of standard components which are used on most projects
- production of components and sub-assemblies to achieve 'right first time' quality
- management of the delivery of components and sub-assemblies to site exactly when needed
- measurement of the performance of completed components and systems
- learning from experience about product performance and durability
- innovation in the design of components to improve construction products.

Sustained improvement

We can have 'ad hoc' improvement and implement this as we go along. This would involve buying new products off the shelf as they became

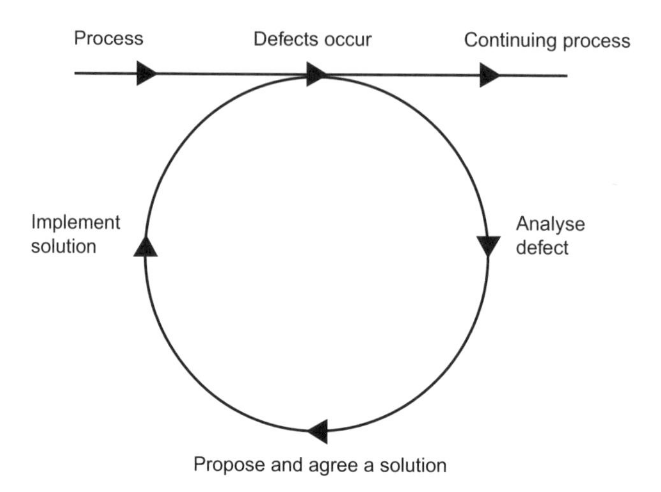

Figure 1.1. Cycle of improvement

available, for example. Or we can galvanise the entire team to be looking at every element of the construction process, its suppliers, its materials, its methods, and the team, and seeking means of improvement on a continuing basis. This is more dynamic. It requires committed leadership and an enthusiastic team. And it must be continuous.

We need also to remain aware of the cycle of improvement which we will be implementing (Figure 1.1).

Lean thinking

As solutions are implemented, the number of defects will decrease. The process continues until a new defect occurs and the cycle is repeated to eliminate the new defect.

'Lean thinking' is about the progressive reduction of waste whenever it occurs. Waste is an activity which does not add value to the product which the customer desires. Elimination, and the decrease in cost arising, gives increased value.

What is lean thinking?

Lean production is the generic version of the Toyota Production System, recognised as the most efficient production system in the world today. Lean thinking describes the core principles underlying this system that can also be applied to every business activity — from designing new products and working with suppliers to processing orders from customers.

The starting point is to recognise that only a small fraction of the total time and effort in any organisation actually adds *value* for the end customer. By clearly defining value for a specific production or service from the end customer's perspective, all non-value activities, often as much as 95% of the total, can be targeted for removal step by step.

Few products or services are provided by one organisation alone, so waste removal has to be pursued throughout the whole *value stream* — the entire set of activities across all firms involved in jointly delivering the product or service. New relationships are required to eliminate inter-firm waste and to manage the value stream as a whole.

Instead of managing the workload through successive departments, processes are reorganised so that the product or design flows through all the value adding steps without interruption, using the toolbox of lean techniques to successively remove the obstacle to *flow*. Activities across each firm are synchronised by *pulling* the product or design from upstream steps just when required in time to meet the demand from the end customer.

Removing wasted time and effort represents the biggest opportunity for performance improvement. Creating flow and pull starts with radically reorganising individual process steps, but the gains become truly significant as all the steps link together. As this happens, more and more layers of waste become visible and the process continues towards the theoretical end point of *perfection*, where every asset and every action adds value for the end customer. Lean thinking represents a path of sustained performance improvement and not a one-off programme.

Applying lean thinking in construction

Pacific Contracting of San Francisco, a specialist cladding and roofing contractor, have used the principles of *lean thinking* to increase their annual turnover by 20% in 18 months with the same number of staff. The key to this success was improvement of the design and procurement processes in order to facilitate construction on site, investing in the front end of projects to reduce costs and construction times. They identified two major problems to achieving flow in the whole construction process — inefficient supply of materials which prevented site operations from flowing smoothly, and poor design information from the prime contractor which frequently resulted in a large amount of redesign work.

To tackle these problems, Pacific Contracting combined more efficient use of technology with tools for improving planning of construction processes. They use a computerised three-dimensional design system to provide a better, faster method of redesign that leads to better construction information. Their design system provides a range of benefits, including isometric drawings of components and interfaces, fit coordination, planning of construction methods, motivation of work crews through visualisation, first run tests of construction sequences and virtual walk-throughs of the product. They also use a process-planning tool known as *Last Planner*, developed by Glen Ballard of the Lean Construction Institute, to improve the flow of work on-site through reducing constraints such as lack of materials or labour.

The *Neenan Company*, a design and build firm, is one of the most successful and fastest growing construction companies in Colorado. The firm has worked to understand the principles of lean thinking and looks for applications to its business, using 'Study Action Teams' of employees to rethink the way they work. Neenan's have reduced project times and costs by up to 30%, through development such as:

- improving the flow of work on site by defining units of production and using tools such as visual control of processes
- using dedicated design teams working exclusively on one design from beginning to end and developing a tool known as 'Schematic Design in a Day' to dramatically speed up the design process
- innovating in design and assembly, for example through the use of prefabricated brick infill panels manufactured off-site and pre-assembled atrium roofs lifted into place
- supporting sub-contractors in developing tools for improving processes.

Enabling improvement

Change is needed if we are to become a truly modern construction industry. The report suggests changes which include improvements in working conditions, skills and training, approaches to design, the use of technology and the relationship between companies.

We should start by valuing our people. It is not just about the quality of our people, it is about the way they are treated. The Task Force view is that the workforce is undervalued, under resourced and often treated as a commodity rather than the industry's single most important asset.

We found people produced better quality and higher outputs when we paid them well and worked sensible hours rather than when we worked for longer hours for less pay.

A Trade Unionist after-dinner speaker made the point that the problems of British Industry were due to workers being overpaid. He belonged to the textile industry. An industry noted for low wages, low levels of training, lack of investment and no apparent strategic plan. An industry where there has been no industrial unrest for decades and annual increases have been amicably agreed. An industry which has nonetheless 'gone to the wall'.

An agreed strategy for relevant investment in plant and an increasing quality at all levels (including the workforce) would have been a sensible alternative.

We often failed to provide suitable site clothing, especially that for use in adverse weather. This was foolish. A little extra spent on protective clothing would be amply repaid. People would be proud of their 'uniform' and team spirit would have improved further. They would also be more effectively protected from the elements, more comfortable and so could have worked better. The better items would probably have lasted longer.

At a time when the UK motorcar industry was in the doldrums and Japanese industry was riding high, a report stated that small tool investment per worker was £20 in the UK and £200 in Japan. This could have been a key issue in our failure to perform.

In our experience, successful jobs were ones where the workforce was provided with fully adequate plant, chosen to maximise performance. Where we failed to achieve this for any reason, then contract performance deteriorated.

Decent working conditions

Site conditions are considered appalling. Good site facilities, and proper uniforms, would help morale and image. If you are treated well and look good, you might perform well. You are very unlikely to perform well under opposite circumstances.

The Egan Report says that most accidents seem to occur when people are not properly trained or are working out of process. The introduction of mandatory training and changes to the procurement process occasioned by changing, or new regulations, should secure an increase in safety standards. Accidents happen because we have not provided a design that is based on maximum safety, where the site is not as safe as it could be, or where operatives are not equipped suitably, or fail to use the equipment provided correctly. Using fast construction programmes in inappropriate conditions does not help either.

There is a link between increasing safety standards and improving performance. You cannot have one without the other. Operating in a role where an overview of many businesses is obtained, we feel that organisations with good safety standards perform better than those with lower standards. A business that has a proper safety culture, one which is embraced from top to bottom of the organisation, will perform more capably than a similar business that relies on its internal enforcement procedures.

More and better training

The Task Force recommends the following.

(*a*) At the *top management* level, there is a shortage of people with the commitment to being best in class and with the right balance of technical and leadership skills to manage their businesses accordingly. The industry needs to create the necessary career structure to develop more leaders of excellence.

(*b*) At the *project manager* level, we see a need for training in integrating projects and leading performance improvement, from conception to final delivery. We invite training organisations, including the professional institutions, to develop the necessary training programmes.

(*c*) The key grade on site is the *supervisor*. The UK has one of the highest levels of supervision on site internationally but one of the poorest records of training for supervisors. We invite the CITB and other relevant National Training Organisations to consider this issue as a matter of urgency.

(*d*) Among *designers* the high standards of professional competence achieved in their training and development need to be matched by a more practical understanding of the needs of clients and of the industry more generally. They need to develop greater understanding of how they can contribute value in the project process and the supply chain.

(*e*) There is not enough *multi-skilling*. The experience of other industries is that heavily compartmentalised, specialist operations detract from overall efficiency. Modern building techniques require fewer specialist craftsmen but more workers able to undertake a range of functions based around processes rather than trade skills. This is being addressed by overseas companies but the UK is in danger of being left behind.

(*f*) Upgrading, retraining and *continuous learning* are not part of construction's current vocabulary. There is already frustration among component suppliers that their innovations are blocked because construction workers cannot cope with the new technologies that they are making available. This has to change.

Our experience concerns us more with *relevance*. We can spend a lot of money wastefully providing training which the business does not need. Individual and business appraisals help focus on the *relevant* needs.

We also tend to use the word 'training' rather loosely. 'Training' is about doing the current job better. Yet much of what we call 'training' is education because it offers a wider knowledge than the delegates already have.

Just as a business becomes more flexible by using sub-contractors to accommodate a varying demand, so it makes equal sense for the individual to obtain a wider range of skills to meet similar variations.

The ICE topics (ICE, 2001) for the Professional Review include consideration of the engineer as a generalist or a specialist. There is a need for both.

The statement that the designer's high level of professional skills should be matched by a more practical understanding of the needs of clients and the industry is correct in some instances. More fundamental is the need to get a greater understanding of how they can contribute value in the project process and the supply chain. In other words, what can a design do to make the task of everyone in the supply chain easier? Everyone includes each site worker, each supplier and each manger throughout the chain. The designer can only learn this from practical experience or by using feedback that will inform that designer of the experiences, good and bad, of others.

Cost pressures within the industry tend to force designers to employ new graduates on design preparation from the start of their employment. Without the knowledge gained from practical experience or from feedback they cannot simplify the work of other members of the supply chain, especially those involved in site construction work.

Mr Avijit Maitra, Chairman of the Reliability Forum, speaking at an ICE meeting in July 2001 stated that 'too many designs are done today by people who do not know what they are doing or why they are doing it'. Harsh words. Yet if a design engineer is lacking in the *relevant* construction experience pertinent to the job, or is not receiving feedback from the supply chain, how can that engineer know — or be expected to know?

Mr Alan Pemberton, a director of the Brian Clancy Partnership, writing in *New Civil Engineer* (2000), felt that the Construction (Design and Management) 1994 Regulations (CDM) have made no difference to safe design. He suggests that, while engineers do not show a reckless attitude to design, few make significant changes to their design in order to achieve a structure that is significantly safer to build, maintain or demolish.

The issue of upgrading, retraining and continuous learning is one of common sense. Any business needs to get the best from its resources and one way to achieve this is to improve the quality of resource. Upgrading and retraining are usually decided by the business needs of the company. Continuous learning can be provided at any time, often at little cost, and usually produces good spin-off results or better company spirit and more interest in the business, a feeling of people being interested in you. These all become available when you use continuous learning.

Examples of effective, low-cost continuous learning include:

- 'toolbox talks,' often on site, given by site staff — they can cover any topic relevant to the business
- office briefing — keep people in the picture
- assignments — transfer of employees to other points of the business to learn new skills, for example put a buyer in estimating or a designer with the resident engineer team
- encourage team working — this helps knowledge to 'rub-off' from one person to another
- internal training sessions — often educational. Use in-house providers — no additional costs are incurred. Delivery must be effective or it will be a waste of time.

The report shows that quality and training are inextricably linked. You cannot have one without the other. It is necessary to educate the workforce, not just in the required skills, but also in teamwork for the required improvement to occur. Preference should be given to contractors who do train workers.

Design for construction and use

The Egan Report states:

- too much time and effort is spent on construction sites, trying to make designs work in practice. How can a business which produces such design data claim to have any knowledge of efficiency, safety or practicality?
- too many buildings perform poorly in terms of:
  - flexibility in use
  - operating and maintenance costs
  - sustainability.

If we are to properly design for construction and use, action needs to be taken to address these issues. It is suggested that:

- *suppliers and sub-contractors* have to be fully involved in the design team. In the manufacturing industry, the concept of 'design for manufacture' is a vital part of delivering efficiency and quality, and construction needs to develop an equivalent concept of 'design for construction'
- the *experience of completed projects* must be fed into the next one. With some exceptions the industry has little expertise in this area. There are significant gains to be made from understanding client satisfaction and capturing technical information, such as the effectiveness of control systems or the durability of components
- *quality* must be fundamental to the design process. Defects and snagging needs to be designed out on the computer before work starts on site. 'Right first time' means designing buildings and their components so that they cannot be wrong
- *designers* should work in close collaboration with the other participants in the project process. They must understand more clearly how components are manufactured and assembled, and how their creative and analytical skills can be used to best effect in the process as a whole. There is no longer a place for a regime of design fees based on a percentage of the costs of a project, which offers little incentive to build efficiently
- design needs to encompass *whole-life costs*, including costs of energy consumption and maintenance costs. Sustainability is equally important. Increasingly, clients take the view that construction should be designed and costed as a total package including costs in use and final decommissioning
- *clients* too must accept their responsibilities for effective design. Too often they are impatient to get their project on site the day after planning consent is obtained. The industry must help clients to understand the need for resources to be concentrated up-front on projects if greater efficiency and quality are to be delivered.

The involvement of suppliers and sub-contractors enables their knowledge to be fully inputted. Many businesses struggle because their staff do not understand those elements of the project. If contract staff are lacking in knowledge then it is reasonable to assume that some designers will be similarly ill informed. Involvement means participation. Participation of people with knowledge means things are more likely to be done correctly.

We found the experience gained from completed projects to be extremely valuable. We got a good idea of how to do things properly, the problems which might occur and how to avoid or get over them. And we performed knowledgeably without the need for great learning curves. Many young designers suggest that project meetings which provide feedback on required issues would help them.

In a discussion on quality, we agreed the following.

(*a*) A quality product in construction is very likely to be cheaper than a lesser product. We gain from things such as 'right first time', a more able workforce is needed to provide quality, and so on.
(*b*) If quality is included in the design, if that design provides better components and makes the project more easily achievable, then defects will be lessened. Clients will be pleased to see better quality workmanship and improved construction performance.
(*c*) By going for quality from the onset, the entire supply chain stands to benefit.

A team will usually work better than a series of individuals. By working as a team the individual skills available can be used as and where necessary to get the best result. If we take this point into the design process, what benefits can we achieve?

It is logical to aim for 'best practice' throughout any project. The designer, searching for 'best practice', is more likely to achieve this by trawling through the skills of others involved in the project and using these, where relevant, to best effect. None of us knows everything. It is logical to collaborate if we want the best results. If you are uncertain about this suggestion, ask 'does anyone feel that we get better results by non-collaboration?'. We suspect that there would be few takers.

The Egan Report comments strongly on the tendency of clients to consider projects on the initial price rather than the final cost. Final cost is the sum expended to the point where the project is replaced. Maintenance and demolition costs can be, and usually are, fully comparable to the initial project cost. It would be beneficial to consider costs incurred over the life of a structure and opt for that which gives best value. Some items will cost more than others. If they are far more durable, however, the extra cost could be very worthwhile. Mechanical equipment may be selected on the basis of reliability and effectiveness, some glazing options will be demonstrably better than others and so on.

Size matters. Plant rooms, lifts, rooms and the buildings themselves often appear barely adequate for today's needs. How can they be suitable for the needs of tomorrow? We have all experienced problems in the use of items that are barely adequate for the job proposed — and regretted not paying a bit more for something much better.

Many contractors complain about the desire of clients to start jobs as soon as the contract is signed. The contractors know that a better job will result if proper planning (and design if necessary) is carried out and preferably completed before a start is made on site. They also know that it is in the client's interest to allow this to happen. Clients need to be made aware also.

Standardisation

The complexity of construction would be eased by the greater use of standardised components. Standardising processes also helps construction teams embark positively on new projects — they have done it before! Experience indicates that very attractive buildings can be designed using a high degree of standardisation.

The scope of standardisation

The Construction Confederation in its evidence to the Task Force told us there was scope to standardise many construction products and components. Examples include:

- manhole covers — local authorities have more than 30 different specifications for standard manhole covers
- doors — hundreds of combinations of size, veneer and ironmongery exist
- motorway bridges — many UK bridges are prototypes, whereas they are of standard construction in France, Germany and Belgium
- toilet pans — there are 150 different types in the UK but only six in the USA
- lift cars — although standard products are available, designers almost invariably wish to customise these.

The Confederation cites the benefits of standardisation as being: reductions in manufacturing costs; fewer interface and tolerance problems; shorter construction periods, and more efficient research and development of components.

Technology as a tool

The use of technology alone will probably fail to increase efficiency and quality. Experience in other industries recommends that we get the culture right, then the process, then apply technology as a supporting tool.

Information technology is useful in the design of buildings and their components and for exchanging information rapidly on design changes. Redesign should take place on computer, not on the construction site.

Too many projects suffer programme and budget overruns because the design is not quite right or the design was not quite ready when work started. Whatever the reason, and whoever pays the extra cost, the bulk of this probably achieves very little in return. There is little wonder that contractors seek to recover such extra cost and clients are reluctant to pay monies for which they receive little in return.

Better regulation

The Task Force view is that, while regulation is necessary, the application and interpretation of regulations varies from area to area. This makes it more difficult to implement projects speedily and efficiently. The variability of enforcement leads to significant extra cost.

There is felt to be scope for improvement by:

- making building control more output driven, enabling project delivery to performance standards rather than detailed prescription
- making the land use planning system more predicable — this would help improve the efficiency of design, particularly of houses.

Long-term relationships

Long-term relationships have been found to be essential ingredients to the delivery of radical performance improvement in other industries. It is suggested that long-term alliances are formed which include all those involved in the construction process.

The Whitbread Hotel Company rationalised its supply chain from 30 contractors to five and embarked on long-term partnering arrangements. Working on the basis of mutual interest, a construction strategy, objectives and improvement targets are set through negotiation between Whitbread, its partners and the supply chain. Whitbread shares its five-year business plan with its partners so that they contribute proactively to the achievement of Whitbread's objectives while planning their own businesses with greater effectiveness. Whitbread agrees fixed amounts for contractors' profits and overheads and shares savings from performance improvement with its partners. Competition within the supply chain focuses upon delivering continually improving performance.

Many major retailers have rationalised their supply chain of contractors. When top retail clients who we were working for rationalised their supply chains we certainly benefited. We saw that we were appreciated and tried harder, the workforce became much more aware of the client's requirements and was able to work more effectively, administration became easier and management was much more knowledgeable. We also became very much a part of the client team.

In this connection, the Task Force wishes to see the following.

(*a*) *New criteria for the selection of partners*. This is not about lowest price, but ultimately about best overall value for money. Partnering implies selection on the basis of attitude to teamworking, ability to innovate and to offer efficient solutions. We think that it offers a much more satisfying role for most people engaged in construction.

(*b*) All the players in the team *sharing in success* in line with the value that they add for the client. Clients should not take all the benefits: we want to see proper incentive arrangements to enable cost savings to be shared and all members of the team making fair and reasonable returns.

(*c*) *An end to reliance on contracts*. Effective partnering does not rest on contracts. Contracts can add significantly to the cost of a project and often add no value for the client. If the relationship between a constructor and employer is soundly based and the parties recognise their mutual interdependence, then formal contract documents should gradually become obsolete. The construction industry may find this revolutionary. So did the motor industry, but we have seen non-contractually based relationships between Nissan and its 130 principal suppliers and we know they work.

(*d*) The introduction of *performance measurement* and competition against clear targets for improvement, in terms of quality, timeliness and cost, as the principal means of sustaining and bringing discipline to the relationships between clients, project teams and their suppliers. The evidence we have seen is that these relationships, when conducted properly, are much more demanding and rewarding than those based on competitive tendering. There are important issues here, particularly for the public sector.

Replacing contracts with performance measurement

Nissan UK and Tallent Engineering Ltd have no formal contract beyond an annual negotiation of the cost and quality of the rear axles that Tallent produce for Nissan's cars, and rigorous targets for improving performance. Each morning Tallent receives an order from Nissan detailing the precise mix of axles required by Nissan and fives times a day Tallent deliver to Nissan's Sunderland plant. If a problem was to occur with quality, Tallent would send engineers to Nissan to fix it on the car production line. If a problem resulted in a significant loss of production, Nissan would expect to compensate Tallen for lost business or vice versa, but this has never happened and both sides work hard to ensure it cannot. Both Nissan and Tallent use similar no-contracts relationships with the firms delivering their construction projects.

Nissan's QCDDM supply chain management system is acknowledged to be among the most effective in the world. It measures all suppliers on **Q**uality, **C**ost, **D**elivery, **D**esign and **M**anagement against negotiated continuous improvement targets. For each element the supplier is marked on a range of product and process items which are aggregated on a weighted basis to give a performance percentage for that element. Competition is created across the supply chain by collating the performance information every month and informing each supplier of its performance in relation to the others.

Reduced reliance on tendering

The submission of a tender is expensive. An achieved rate of one success for each six tenders submitted is regarded as quite good. Nonetheless, it means that effort put into five unsuccessful efforts is largely wasted.

In times of recession the success rate may be only one in twelve. At a time when the reduction of cost is vital, we are wasting even more on the tender process.

Contractors will be glad to rely less on tendering, not just because of the likely reduction in abortive costs, but because of the 'Dutch Auction' which results from the acceptance of the lowest price.

The Task Force believes that with:

- quantitative targets
- open book accounting
- a demanding selection process for partners

then value for money can be adequately demonstrated and properly audited.

Summary

The key issues of the two reports can be summarised. *Constructing the team* by Sir Michael Latham (1994) states that:

- clients have a role in promoting good design which provides value for money in both total cost and cost in use — energy and maintenance costs are of prime importance, together with appearance and the effective use of space (section 1.18)
- the procurement route taken by a client should be judged on the basis of the type of project and the amount of risk the client wishes to take (section 3.8)
- only when the client finalises his ideas on project need, the risk to be taken and his own involvement, should the project and design briefs be prepared
- effective design is crucial — effective design demands leadership, coordination of effort, everything covered, a well-informed client, no 'fuzzy edges' between parts, no deficiencies in the information provided, signing-off work for correctness as work is completed (section 4.1)
- a modern contract should include (section 5.18):
  - a specific duty for all parties to deal fairly with each other
  - firm duties of teamwork so all can share in success
  - a clear definition of the roles and responsibilities of all parties
  - easily understood language with guidance notes
  - a choice of risk allocation
  - avoid variations — if you must have them, price them in advance
  - ensure fair, prompt and secure payment to all parties
  - avoid conflict on site and, where dispute occurs, resolve it speedily
- when selecting suppliers or tenderers, consideration should be given to (section 6.1):
  - quality as well as price
  - a lead manager

    - sensible tender lists
    - no burdensome qualification procedures
    - value for money
    - cost in use
- factors which determine performance include:
    - adequacy of safety consideration
    - ability of operatives
    - improved training
    - relevant professional education
    - adequate research and development which is properly funded (see chapter 7 of the report)
- teamwork on site depends on fair treatment for all parties (see chapter 8 of the report)
- dispute resolution should be speedy and fair (see chapter 9 of the report)
- payments should be secure and protected against insolvency of a party (see chapter 10 of the report)
- the liability for defects post-completion should be properly addressed (see chapter 11 of the report).

Rethinking construction by Sir John Egan (1998) states that:

- the UK construction industry needs to improve — there should be:
    - less fragmentation
    - increased profitability
    - more capital investment
    - more investment in R&D
    - more and better training
    - differentiation between price and cost
- we should set targets for improvement and give commitment to the needs of the customer, to quality, and to people — the team and the process should be integrated around the product, and management should be committed to improvement throughout the organisation
- the construction process should be improved by:
    - more process repetition on projects
    - integrating all team members in a manner which uses their full range of skills to deliver value to the client
    - focus on the end product — how best to achieve the desired result
    - continuous product improvement

  - implement projects correctly, suiting the site and the customer
  - partnering throughout the supply chain
  - once improvement starts, maintain it
- we can get improvement by developing a better culture in the industry:
  - providing decent working conditions
  - more and better training
  - produce designs which consider the needs of construction and project end use
  - standardise components where possible
  - improve regulatory controls
  - implement long-term relationships.

Chapter 2

The health and safety issues

The Health and Safety at Work Act 1974

The Health and Safety at Work Act 1974 provides a comprehensive framework of legislation. The intention is to promote high standards of health and safety in the workplace. Its aim is to promote an awareness of safety and effective safety standards in every organisation.

The Act aims to involve everyone in all matters of health and safety. Employers, employees and self-employed, the manufacturers of plant, equipment and materials, and those who control premises are involved, together with employee's representatives. With minor exceptions, all persons at work are covered.

The Act is an enabling measure, superimposed on existing health and safety legislation. Regulations made under the Act impose a statutory requirement and the relevant statutory provisions include, in Section 15 of the Act, all Health and Safety Regulations since 1974.

Some of the broad intentions of the Act are as follows.

(*a*) *Maintain a safe site* — one that is without danger to the general public, employed persons or the self employed. Everyone on site must consider the health and safety of others on that site.

Experience shows that sites which are untidy, inefficient or have slack controls tend to be unsafe. These sites are often loss making. On the other hand, tidy and efficient sites tend to be safe and profitable. Safety pays dividends.

Imagine yourself working in unsafe conditions. Would your production be affected by those conditions? Of course it would — in a negative manner.

(*b*) *Provide and maintain plant and systems of work that are safe and without risk to health* — plant and equipment which are able to do the job safely and are adequate for the task. Plant that will do the job properly without overstretching its capacity, breaking down unexpectedly, or suffering from parts failure.

Fully adequate plant may cost a little more than plant which is not quite as good — is less powerful, older, less well-maintained, for example. However, for the small increase in cost you will get much more effective performance. The cost increase is a worthwhile investment.

(*c*) *Ensure safety and absence of risks in the use, handling, storage and transport of articles and substances.* The items we use in our structures are required to be safe in themselves, and stored and moved around safely.

This will enable us to do the job more easily (or efficiently), will usually reduce waste by ensuring better planning, and will ensure the provision of materials which, by being safer, will be able to be used more effectively. The regulation covering the Control of Substances Hazardous to Health (COSHH) arises from this provision.

(*d*) *Provide adequate instruction, and training and supervision as is necessary, to ensure the health and safety of employees.* A workforce knowing what to do and being able to do it safely will out perform a force which is not capable.

(*e*) *Provide a safe place of work, with safe access and egress — provide a safe working environment that is without risks to health.*

There can be many places of work on a project and some of these will be changing regularly. The duty to provide safe access to and from these working places is best served by careful initial planning of how safe access and egress will be achieved — then providing this at the onset of construction. Get it right first time. You will find this is much more efficient than not quite getting it right and then making changes. Getting it right first time also tends to set the example for other areas. The workforce and its plant moves more confidently and works more competently.

The Management of Health and Safety at Work Regulations 1992

Regulation 3 of the Management of Health and Safety at Work Regulations 1992 introduced the principal of 'risk assessment' in that every employer and self-employed person shall make a suitable and sufficient assessment of:

- the risks to the health and safety of employees or themselves to which they are exposed while they are at work
- the risks to the health and safety of persons not in his employment arising out of, or in connection with, the conduct by him of his undertaking.

The intention is to:

- note any hazards
- assess the risks arising from the hazards
- as far as practical, eliminate the risk by removing the hazard
- if you cannot eliminate the risk do what you can to reduce it
- draw the attention of others to the risks that remain.

The introduction of this regulation meant that contractors started to reduce risks on their sites in a systematic manner. However, contractors still had to work to construction data to which the regulation did not apply. There was a limit to which risks could be removed or reduced.

The Provision and Use of Work Equipment Regulations 1992

In general terms, the provisions of the regulations require every employer to ensure the following.

(*a*) The work equipment provided is suitable for the purpose it was provided for, i.e.:

  (*i*) It is only used for the intended purpose. That it is not misused. At the same time there is an implication that appropriate equipment be supplied for each purpose.

  The labour on any project has a certain cost and that cost is best justified by ensuring optimum output. The provision of fully appropriate work equipment and ensuring that it is properly used will help ensure that output is optimised.

  (*ii*) It is properly maintained. We have had dealings with a number of plant operations. The most successful by far was one where we carefully selected the plant needed, ensuring high utilisation, charged it to sites at proper rates and maintained it carefully. As a result, plant lasted 50% longer and profits were good.

(*iii*) Records are kept of its use and maintenance. Records are essential if we are to understand the situation correctly (feedback).

(*iv*) It is only used by approved people. The intention is to prevent accidents by ensuring that unapproved people do not use any item. There is a further saving, however. Unapproved usage often leads to misuse, damage and increased cost. This should now be avoided.

(*v*) Controls are fitted. We can rapidly conclude that any item, fitted with the appropriate safety controls, will be more confidently used than items without such control.

(*vi*) Safe systems of work are implemented. This is more efficient than unsafe systems! Get the correct items, use and maintain them properly keeping the appropriate records. Make sure you have proper safety controls fitted and only allow authorised people to use them.

(*b*) Persons required to use the equipment are adequately trained. Training will improve ability and hence efficiency as well as safety.

The Manual Handling Operations Regulations 1992

Employers shall, so far as is reasonably practicable:

- avoid the need for employees to undertake any manual handling operations at work which involve a risk to their being injured
- where it is not possible to avoid the need to do risky handling, the employer should carry out a risk assessment
- employees shall make full and proper use of any equipment provided to give help with handling.

Far too many people are injured by inappropriate lifting. It is sensible to change the design of the product to reduce excessive weights. More sensible weights not only reduce injuries, they make the job easier. We know of jobs which were made difficult to execute because the manual lifts were too heavy.

Reinforcement, dense blockwork, the lifting of weighty items repetitively by bending the back, are all examples of lifting which can cause injury. Inappropriate weights of the items involved will also slow the work down, cause programme overrun, and increase costs.

Personal Protective Equipment at Work Regulations 1992

Every employer must make suitable and adequate personal protective equipment available and:

- ensure that it is used properly
- ensure it is maintained and stored properly
- give adequate training in personal protective equipment use and the hazards which cause it to be used.

The Egan Report (1998) recommends the use of uniforms. Provide outdoor weather protection clothing of a good quality. It is more comfortable, easier to work in, and lasts far longer. It is less likely to be cast aside.

Health and Safety (Display Screen Equipment) Regulations 1992

Every employer shall:

- carry out a risk assessment to see what risk operators are exposed to
- ensure work stations are suitable for use
- provide eye tests on request
- provide adequate health and safety training in workstation use — eyesight is precious, take care of it — ensure you use display screens in an efficient way
- provide health and safety information.

Workplace (Health, Safety and Welfare) Regulations 1992

Every employer must do the following.

(*a*) Maintain the workplace, equipment, devices and systems in an efficient state, efficient working order and in good repair. A stated example of the direct link between efficiency and health and safety
(*b*) Provide every enclosed workplace with sufficient ventilation of fresh or purified air. Enclosed workplaces occur more often than we realise. Not only can they be unsafe, their situation — size, location, surroundings — can prevent efficient working in such places. In terms of work study, the volume of a working area is never sufficient to make ventilation unnecessary.
(*c*) Ensure the workplace temperature in buildings is reasonable. The temperature for efficient working varies as the working condition varies. For sedentary work, a temperature of 20°C to 22°C is usually adequate. For heavy work in a stationery position, the figures are 14°C to 16°C. When workers feel cold, they become less efficient. It is good business practice to provide a sensible working temperature, the correct clothing, and adequate hot drinks which protect workers from dehydration.
(*d*) Provide suitable and sufficient workplace lights. Some 80% of the information required to do a job is reliant on visual perception. Good visibility is essential if we are to do the work more quickly, to reduce defects and prevent visual fatigue and headaches. Inadequate visibility and glare frequently cause accidents.
(*e*) Keep the workplace clean and remove waste. A tidy site is an efficient site. An untidy site gives not just a poor image, it shows a lack of care. And the workforce stops caring.
(*f*) Provide adequate space and workstations. In construction efficiency terms we need to avoid people working too closely together — they get in each other's way. Tempers rise and efficiency falls when we fail to do this.
(*g*) Ensure floors and traffic routes are suitable for their intended use. The 1974 Act demands the provision of a safe place of work with safe access and egress. Almost the same thing *but* traffic routes suitable for their intended use. Every construction site has its traffic routes and sometimes these are not planned, built or maintained adequately. The end result is that they are less efficient in use and this must add cost to every operation which uses a route, each time the route is used. Get it right first time — and make it safer.
(*h*) Protect people from falls and falling objects. Perhaps the best way to look at this is to estimate your personal loss of efficiency, when you have to work on an untidy site — rubble, nails projecting,

things falling, etc. Would you say 30–50%? Whatever figure you suggest, it will be quite high.

(*i*) Mark or provide safety material in doors and windows and enable the safe cleaning of them. The breakage of glass not only causes injuries. In lifelong cost terms it can be expensive. It is efficient, as well as safer, to mark or provide something safer.

A window cleaning company was suffering a loss of trade due to a high injury rate among its cleaners. The management response was to get the workforce together and give them a good 'dressing down'. 'The next man to fall off a ladder and break his ankle will get sacked', said the boss. If that is the depth of thought which goes into window cleaning, then engineers and architects are well advised to give the operation careful consideration.

There are numerous examples of new structures being built which have inadequate consideration for window cleaning (and replacement).

(*j*) Organise access and egress (traffic routes) so that vehicles and pedestrians can circulate in a safe manner. Few motorists enjoy having to negotiate one-way traffic routes. The chaos they would face if such routes were returned to two-way working would clarify their thinking. The quarrying industry goes to some lengths to arranges its traffic flows and tries to avoid vehicles having to reverse.

(*k*) Provide sanitary conveniences, washing facilities, drinking water, clothing stores and changing facilities, rest and eating facilities. People respond positively when good facilities are provided.

Other regulations cover items such as:

- scaffolding
- hoists and hoist towers
- ladders, steps and staging
- noise
- excavations
- asbestos
- lead
- flammable liquids and gases
- fire
- electricity, and so on.

The intention is to provide an all-embracing matrix of regulations which will warn us of the relevant hazards and guide us towards good practice. Site-based management is usually fully conversant with the relevant regulations and able to work in compliance of them on a day-to-day basis.

The fact remains that the regulations prior to 1994 related largely to what happened on site. They did not cover the performance of clients, designers and architects, insofar as what they asked contractors to construct. Two structures can look quite similar, yet one can be detailed in a manner which makes it extremely difficult to construct, the other can be easy to construct. The easy to construct version will have been designed with its buildability in mind. It will probably cost less, be built more quickly, safely and will even be of higher quality than the difficult construction. It is likely that the easy version will have been developed using relevant experience of all aspects of the work, the more difficult project is likely to have suffered due to a lack of such experience in its design team. One example is likely to give 'best value' and demonstrate good practice. The Robens Report, published in 1972, contained the comment.

> The primary responsibility for doing something about the present levels of occupational accidents and disease lies with those who create the risks and those who work with them.

The traditional regulations largely cover those who work with the risks. Clients and designers, in producing their project requirements and data, can create those risks. There was a requirement for further regulation to encourage the provision of projects which were safe to build, and free of risk. This provision was made by the Construction (Design and Management) Regulations 1994 (CDM).

The Construction (Design and Management) Regulations 1994

The CDM Regulations are an Approved Code of Practice (ACOP), which sets out the way construction work covered by the regulations should be designed and managed from concept to project completion and then on through the maintenance, alteration and demolition phases. That is, they cover the lifespan of a project. All but the smallest of projects are covered by the regulations.

The regulations are additional to, and do not replace, any existing regulations. They are made under the provisions of Section 15 of the

Health and Safety at Work Act 1974. They must be read and given effect with the provisions of the 1974 Act and the 'six-pack' of 1992 regulations.

There are legal duties to be met by all employers, the self-employed and by those not employed but who control persons at work, who thus have responsibilities under CDM when they take decisions which affect the project. Risk assessment becomes mandatory and needs to follow the principles of prevention and protection. In general terms, we try and remove hazards to prevent risk occurring. If this is not possible, a hierarchy of prevention measures is defined.

Designers, as employers or self-employed persons, have a duty to assess the risks which arise from their undertaking. The regulations clarify the requirement on designers to follow certain principles when they carry out their design work, in order to help protect people undertaking the construction work. Key definitions in the CDM Regulations are as follows.

(*a*) *Competence*. Everyone shall ensure that any person to be appointed is competent insofar as health and safety is concerned. We cannot appoint anyone who is incompetent and must show competence in our own efforts.

People or organisations who have no experience of a particular type of project may not be competent to design or construct it under the terms of CDM.

This should ensure that we employ people, or organisations, who are experienced in the relevant field. We will be more likely to have a successful project if we do this.

Relevant skills are required and they need to be available at the right place at the right time and for the required period. A regular complaint from local authorities in Scotland is that English contractors put forward proposals for tender selection. These proposals list people likely to be on site. The contractor gets the job and different staff appear — not a very satisfactory situation. While this does not mean the job will not go so well, it does not offer the best starting point for a cooperative venture.

(*b*) *Adequacy*. The resources allocated to the work shall be adequate insofar as health and safety is concerned. This means that the right amounts of labour and plant are provided to carry out the construction work correctly. It also means that adequate resources will be provided at the design stage, for maintenance or alterations, and for demolitions.

(*c*) *Resources*. Paragraph 36 in Regulation 9 of CDM states:

> Resources is a general term which includes the necessary plant, machinery, technical facilities, trained personnel and time to fulfil the obligations.

In summary, we have, in terms of health and safety, to ensure competency in those we appoint, and adequacy of resource. As the health and safety issue is fundamentally affected by what happens on site it means that:

- the appropriate amount of time is allowed to carry out each phase of a project, from concept to demolition. Only with proper time can designers do a full and appropriate design, can contractors prepare their plans to do a good job and have a sensible programme. To deny time is to rush things. When we rush we get things wrong. People get injured. Provision of the appropriate time must help buildability
- trained personnel, to be effective, need to be trained in the skills required for the particular project. It is likely to mean more effective teams at all levels. Designers and project planners so trained are likely to produce better plans and details than those not trained relevantly. Not only will health and safety improve, but so will buildability
- competency of all involved, Each person is appointed by someone. This should avoid the 'cowboy' element and give better opportunities to those who are professionals in the fullest sense. Quality, teamwork and efficiency are likely to improve
- adequacy of resource of trained personnel and plant must also mean the correct numbers and types to do the work. A concrete gang requires a certain strength to do a particular task. Reduce that strength and accidents are likely to happen
- if we have the correct time to carry out any task and use the correct resources, then the correct cost of the task is determined. Clients should be fully aware of the correct cost of doing the job using adequate and competent resources. They should not look at the lowest price submitted and accept the contractor's word that he has the correct allowances for health and safety. There is also a case for the contractor to be asked to demonstrate his resourcing intention by presenting the resourced programme on which his tender is based.

The satisfying of the CDM Regulations in respect of adequacy, competency and resource should not only bring improvement in matters of health and safety. It should also enable the desires expressed in the Latham and Egan reports to be achieved more easily.

The CDM Regulations consider the safety of people working on structures for the lifespan of a project. This includes:

- constructing
- cleaning
- maintaining
- repairing or altering
- demolishing.

The designer, in preparing his design, must prepare it in a manner which considers such safety. The best design will incorporate buildability for the lifespan of the structure. Efficiency is likely to improve and costs reduce on each process carried out.

The CDM Regulations impose duties on clients. These are as follows.

(*a*) Appoint a planning supervisor who is competent and will provide adequate resources. The planning supervisor is responsible for the health and safety aspects of the project. Appointment of the correct planning supervisor will strengthen the abilities of the client team considerably. This should lead to all round benefits.

(*b*) Appoint a principal contractor who is competent and who has adequate relevant resources. Correct appointment should help ensure a smoother running contract and lessen the likelihood of cost overruns.

(*c*) Appoint designers if required.

(*d*) Ensure that those appointed will provide adequate resources and be competent. Capability for the project should thus be assured.

(*e*) Provide adequate information regarding the site or premises. This would include adequate ground investigation, previous use of site, the work processes carried out and any existing drawings or other data, for example.

A senior consultant, when asked to make public comment on ground investigation, stated that too many jobs went wrong because of late or inadequate ground investigation. A sports stadium suffered a £15 million cost overrun when the ground proved different to that envisaged. A sewage works suffered a £0·5 million cost overrun when unexpected contaminants were found buried in the ground. A concrete framed chemical plant was found to be dangerously contaminated when demolished.

The cost of taking the demolition debris to a controlled tip was £750 000 more than anticipated.

(*f*) Control the start of work. The client must ensure that the principal contractor has prepared an adequate health and safety plan for the construction phase before allowing work to commence. This takes time. The contractor needs to decide who his sub-contractors will be, consider their work methods, and then put these into the construction plan.

A complaint heard from contractors is that clients want a quick start on site — or someone else will get the job. This seems poor business sense. Construction work goes much better when properly planned. Orders are best placed prior to entry onto site. Construction programmes need finalising before work starts. The sub-contractors who will be involved initially need to feel part of the team.

(*g*) Where the construction work overlaps the client's own work, control health and safety. The client needs to impose his requirements onto the contractor. This would apply in a retail store or a factory. It should ensure not just the health and safety of those involved but more effective and efficient working.

(*h*) Keep a health and safety file for future use. When the project requires further work — be it extensions, modifications, repair, indeed anything — we will have the full data of the entire structure on which to base our proposals. No assumptions should have to be made as to what is already there.

These duties should not only assist in matters of health and safety — they are measures of common sense. If implemented as intended, considerable benefits can be gained in terms of increased efficiency and buildability.

The CDM Regulations impose duties on designers. These are as follows.

(*a*) Make the client aware of his duties. There is a legal duty to explain. A correctly worded and standard letter from the designer will assist in many cases. It will also be a record for the future. Properly briefed, a realistic client is likely to respond positively and benefit as a result.

(*b*) Design with due regard to health and safety. This is achieved by removing risks altogether, if this is possible. If it is not possible, the risks should be reduced as much as is practicable. It means that design teams need to be familiar with the required construction methods and the later requirements (maintenance, for example) of the project. Be aware of the risks and how best to avoid or reduce them. They need to know the design details which will ensure safe working later during the life of the project. This requires practical experience of the type of project concerned. It is an area where the inexperienced designer has, and always has had, difficulty.

(*c*) Provide adequate information on the remaining risk to those who need it. This will alert those working on a project at any time to the dangers they will face. They can properly plan to deal with the dangers as they carry out the works required. The work can then be carried out more smoothly and efficiently.

In carrying out their role, designers should:

- ensure that the nominated design team is competent and fully aware of its duties — this requires experience of the type of work involved and the methods to be used in carrying it out. This experience will help minimise risk and maximise buildability
- identify project hazards — list the hazards likely to arise. A formalised inventory system helps. Feedback from past jobs can be used to prepare the hazards listed in the inventory
- carry out risk assessment on the hazards — this will identify the level of risk. The higher the risk, then usually the greater the danger
- alter the design to remove the risks where possible
- if the risk cannot be removed, try and reduce it — high risks should be prioritised for removal. It is *not* practicable to remove all risk. The expectation is to remove those which can reasonably be removed
- pass on information about remaining risks — this is via the pre-tender health and safety plan
- cooperate with the planning supervisor and others — this is team working. It includes other designers. It enables overall risk (and construction) considerations to be made. It will help prevent 'fuzzy edges' developing as the Egan Report (1998) recommends. It should help designers to prepare designs that consider the features of the designs of others which are superimposed. The collaboration of structural and services designers can clearly be beneficial.

The Construction (Health, Safety and Welfare) Regulations 1996

The Construction (Health, Safety and Welfare) Regulations 1996 are largely updates of existing requirements. There is, however, a key difference. The earlier regulations were prescriptive — the width of scaffolding required for various uses, the requirements for supporting excavations, were clearly stated. The new regulations place wide-ranging duties on those involved in construction. They say what you must prevent — not how you prevent it. For example:

- Regulation 5 — there is a general duty to provide a safe place of work and safe means of access to and from that place of work
- Regulations 6 and 7 — precautions against falls. Prevent falls from height by physical precautions or, where this is not possible, provide equipment that will arrest falls
- Regulation 8 — falling objects. Where necessary protect people at work and others, and take steps to prevent materials or objects from falling
- Regulations 9, 10 and 11 — work on structures. Prevent accidental collapse of new or existing structures or those under construction
- Regulations 12 and 13 — excavation, cofferdams and caissons. Prevent collapse of ground both in and above excavations
- Regulation 14 — prevention or avoidance of drowning
- Regulations 15, 16 and 17 — traffic routes, vehicles, doors and gates. Ensure construction sites are organised so that vehicles and pedestrians can both move safely and without risks to health
- Regulations 18, 19, 20 and 21 — prevention and control of emergencies. Prevent risk from fire, explosion, flooding and asphyxiation. Provide fire-fighting equipment, alarms and emergency routes. Make arrangements to deal with emergencies
- Regulation 22 — welfare facilities to be provided
- Regulations 23, 24, 25, 26 and 27 — site-wide issues. Provide suitable fresh air, working temperature, lighting, clean sites and so on
- Regulations 28, 29 and 30 — training, inspection and reports.

For the inexperienced, reference to the earlier regulations will help them comply with the 1996 requirements.

Practical application of the regulations will benefit not just health and safety but the whole construction process. While a systematic approach to any topic is desirable, when the system becomes the objective it can become redundant as an effective tool. We need to look further than the regulations themselves to decide how to avoid the risks occurring in the first place. This process must start, as Lord Robens (1972) suggested, with clients and designers. Their primary responsibility is to eliminate, or at least reduce, the risks which they create.

Before anything can be done about risk avoidance, we need to know what accidents occur, why they occur, the number of accidents associated with the various risks and the seriousness of the accidents themselves. This requires the recording of accidents, much of which is mandatory. It also requires experience on the part of the designer as to how accidents and risks can be eliminated or at least reduced.

The experience required comes from practical experience of the construction process relevant to the project or from feedback provided by others as to what goes wrong and why.

Feedback is an excellent tool for getting data from the workplace to management level. Business improvement is achieved to a large degree by the elimination of defects (this requires feedback) or by improving on the strengths of the business. The elimination of defects usually provides the best returns. We should seek feedback and use it to the full.

The publication *Designing for Health and Safety in Construction* (HSE, 1995) provides guidance for designers arising from the introduction of the CDM Regulations. The regulations have an impact on all stages of the planning and management of a project. They cover all designs and specifications for a project and all those involved in their preparation. At the concept stage fundamental decisions are taken about the design and these can affect construction health and safety. The location of a structure on site, the consideration of likely temporary works requirements, the fixing methods, and the solvent content of adhesives, all affect health and safety, hence risk and, quite probably, buildability issues.

The traditional practice has been for designers to leave many of the health and safety risks for contractors to solve, losing the opportunity to reduce risk at the design stage. The first step that designers must now take is to recognise the risks involved in the project they are concerned with. Table 2.1 lists areas where most injuries occur.

Surfacing and groundworks are obvious dangers, producing high injury rates in what are relatively small areas of the industry. The structural steel industry has put a lot of effort into reducing risk. Maintenance is largely carried out by tradesmen doing work which has a high degree of repetition. Could this explain the lower number of accidents? Table 2.2 lists the incidence of fatalities that arise from various types of accidents.

Table 2.1. Areas where most injuries occur

Process	Total of reported non-fatal injuries and (fatalities) 1993/94
Finishing/plastering/glazing	5745
Transfer of people/materials (on site)	2186
Surfacing: paving/road laying	1573
Groundworks: excavation/sewers	1428
Handling	1345
Loading/unloading	1154
Labouring	868
Bricklaying	789
Structural erection	722
Scaffolding	663
Maintaining repair/renovation	434

With regard to falls from height, our designs should consider points such as the following.

(*a*) Minimising the work carried out at heights, e.g.:
- (*i*) increase precasting
- (*ii*) increase speed construction rates by repetitive and simplistic detailing
- (*iii*) use low maintenance external fixtures (gutters, fascias, downpipes)
- (*iv*) design windows which can be cleaned from inside the building, or provide bespoke cleaning systems
- (*v*) minimise the need for scaffolding (speedier construction means a shorter requirement) — take greater care in selecting the scaffold arrangement to be provided
- (*vi*) prefabricate at ground level as much as possible
- (*vii*) make roof construction, including roof lights, gutters and other fittings stronger and more durable — this will reduce risk, increase the life of the roof and reduce maintenance frequency

Table 2.2. Incidence of fatalities that arise from various types of accident (continued overleaf)

Type of accident	No. of fatalities	Percentage of total
Falls from a height		
Ladders (all types)	9	
Scaffolding (all types)	8	
Fragile roofs	8	
Roof edges or holes in roofs	6	
Structural steelwork	4	
Temporary work platform (above ground)	3	
Parts of floors/surfaces not listed above	4	
Other	1	
Total	43	56%
Trapped by something collapsing or overturning		
Buildings/structures (or part of)	7	
Earth, rocks, e.g. trench collapse	3	
Plant including lifting machinery	1	
Scaffolding collapse	1	
Vehicles falling from supports/ overturning	3	
Other	1	
Total	16	21%
Struck by a moving vehicle		
Bulldozer	1	
Excavator	2	
Private vehicle	1	
Road tanker	1	
Trailer	1	
Other	2	
Total	8	10%

Table 2.2. continued

Type of accident	No. of fatalities	Percentage of total
Contact with electricity or an electrical discharge		
Domestic-type equipment	1	
Hand tools or hand lamps	1	
Overhead lines	2	
Total	4	5%
Struck by a falling/flying object during machine lifting of materials		
Total	3	4%
Contact with moving machinery or material being machined		
Conveyor belt		
Hoist		
Total	2	3%
Exposure to a hot or harmful substance		
Total	1	1%

(*viii*) provide proper protection at roof edges and holes in floors and roofs
(*ix*) pre-assemble structural steelwork at ground level
(*x*) provide an erection sequence for structural steelwork which will ensure safety at all times during the erection process — the design will conform to this
(*xi*) site supervision must be adequate and competent.

(*b*) The prevention of people being trapped by something collapsing or overturning. We should consider this on the following:

(*i*) structural alteration work, especially in older properties, needs to consider the possibility of a weakened structure or weakened materials in the structure and provide adequate temporary works to cater for an increased need (partial demolition — knocking holes in walls, can be a particular problem)

(*ii*) site supervision, supported by a design input as necessary, must be adequate

(*iii*) temporary works design should consider not just the immediate construction but the ongoing stability of adjacent parts of the structure

(*iv*) design for safety by using robust and fully adequate replacement parts — steel beams used as lintels over a new opening; adequate sole plates for those lintels; these should allow for the structural deterioration of the material they are seated on, especially if it is masonry or stone of any type

(*v*) use more durable timber to minimise the frequency of maintenance

(*vi*) never undersize new structural elements — allow for future safety and satisfactory performance under changed conditions.

In ground works:

- locate services accurately and design temporary works which take due consideration of each service and its location
- consider using high-level carrier drains with backdrops to deeper sewers — future work will be safer and cheaper by connecting to the carriers
- small underground pumping stations, or similar structures, could be provided as single fully fitted out and operational units — this will minimise construction work below ground and give a better quality product as well as save construction time on site
- design for the future — incorporate future capacity increases in the drains supplied. This will increase the lifespan of a drainage system. A larger pipe will cost very little extra and is likely to be cheaper than the minimum diameter in lifelong cost terms.

In terms of plant and lifting machinery:

- design structural elements, plant and other items to be incorporated in a structure, in a manner which considers safe and simple placing within the structure. Provide lifting points which ensure a level lift rather than a tilted one. Ensure the lifting points are adequately sized. Can they be left in position for future use?

- design plant rooms which are adequate in size to enable safe maintenance of the machinery in them — make them light and airy, and positioned to ensure safe access.

If struck by a moving vehicle:

- the quarrying industry goes to great lengths to eliminate, as far as possible, the reversing of vehicles — this commences at quarry design stage
- future road and bridge designs could consider the separation of vehicles from the workforce (this is considered more closely in Chapter 5)
- minimise road maintenance requirements by improved design.

If in contact with electricity:

- ensure the positions of supplies are carefully recorded and that details are made available to those carrying out electrical work
- ensure all electrical work is approved — protect future occupiers from the DIY current resident.

If there is exposure to hot or harmful substances avoid their use if at all possible — the COSHH regulations have helped in this area.

Table 2.3 provides estimates of the numbers suffering from work-related ill-health. The issue of work-related ill-health can be split into two parts, as follows.

(*a*) What we use:

(*i*) *Asbestos*. Strict regulations cover the use and removal of asbestos material of all types. While conformance will help minimise any future problem, it is essential that the likely presence of asbestos is considered in any existing structure at the preliminary phase of any project.

(*ii*) *Skin disease*. Contact with many types of traditional construction material can lead to dermatic conditions appearing. The COSHH regulations have led to many materials being changed in content or replaced by something better. Contact with wet concrete can create problems. The concrete placers are protected by suitable clothing. Designers can assist by designing concrete mixes which are easy to place in the structure (i.e. easily workable mixes). The use of non-vibrated concrete could be

Table 2.3. Potential number of persons suffering from work-related ill-health (continued overleaf)

Hazard	Possible resulting disease or condition	Estimated lower limit	Estimated upper limit	Ref. source*
Asbestos	Mesothelioma	200–250 deaths†		ARSS
	Asbestosis	100 cases‡		
	Lung cancer	At least 200–250 deaths§		
Musculo-skeletal injury	Total	30 400	48 100	LFS
	Back disorders	14 700	27 800	
	Work-related upper limb disorders (WRULD)	4600	13 000	
	Lower limb disorders	1100	6500	
	Unspecified musculoskeletal	3800	11 700	
Respiratory disease	Lower respiratory disease (bronchitis, emphysema, etc.)	2000	8500	LFS
	Asthma Pneumoconiosis (excluding asbestos)	1000	5800	
	Upper respiratory disease (sinusitis influenza, etc.)	300	4400	
Skin disease	Dermatic conditions	3100	10 500	LFS
Noise	Deafness and ear conditions (tinnitus)	1000	5800	LFS

Table 2.3. continued

Hazard	Possible resulting disease or condition	Estimated lower limit	Estimated upper limit	Ref. source*
Ionising radiations	Radiation exposure	50 (exposed to more than 15 mSv)		CIDI ARSS
Lead	Lead poisoning	18 (above suspension level of 69 ug/100 ml)		ARSS
Compressed air	Decompression sickness	50		HSE

*Reference source list:

ARSS 1993/94 HSC Annual Report Statistical Supplement
LFS 1990 Labour Force Survey
CIDI 1986–1991 Analysis of doses report to the Health and Safety Executive (HSE) Central Index of Dose Information
HSE HSE estimate (from HSE tunnelling expert)

Note: Evidence from the supplementary 'Trailer' questionnaire to the Employment Department 1990 LFS indicates that the true annual prevalence for occupational ill-health categories identified above among people working in construction operations during the three years prior to spring 1990 is likely to lie between the lower and upper limits quoted.

† Based on last full-time occupation recorded on 1991 death certificates.
‡ Same proportion as above applied to disablement benefit cases.
§ Based on mesothelioma death certificate proportion.

considered more widely. Increased use of precast products may help. Repetitive and simple detailing of a structure will assist concrete placing.

(*iii*) *Respiratory disease*. The creation of concrete or other dust should be minimised by providing suitable ducting in the initial design. This will avoid the need to cut channels in finished work.

Minimise the need to cut materials to size by standardisation. Then order the appropriate correct sizes.

The COSHH regulations have had a positive effect on the materials we use. There is now a much wider awareness of the dangers inherent in the use of various materials and the need for individuals to protect themselves when using these materials. There is an awareness that prolonged use can create problems. Spirit-based materials, glues, liquids and paints have been redesigned to minimise the risk arising from their use.

There is an awareness that the compounds used in production of older types of artificial stone can create problems.

All products cut on-site can give problems if exposure to the arisings is prolonged. Protection is necessary.

(*b*) How we use it:

(i) *Muscular skeletal injury*. The figures indicate that between 30 000 and 48 000 workers suffer each year. Much of this is due to lifting items which are too heavy and carrying out work in a manner to which the body is unsuited.

Lifting items that are too heavy

The Manual Handling Operations Regulations 1992 require that manual handling be avoided where possible, that risk assessments be carried out on items to be handled and the risk of injury arising from lifting be reduced so far as is reasonable.

Guidance is given on weights which can be sensibly lifted but these vary with the conditions under which lifting is carried out. Certainly 25 kg is a realistic expectation.

The construction industry has introduced a considerable amount of mechanisation to enable safe handling on sites. Factors which designers should sensibly cover in their designs are:

- design concrete to be as workable as the design will permit. This will simplify the effort of placing. It will also speed progress and give a higher quality finish (if the mix is correctly designed). Many concrete gang operatives are 'burnt out' physically before they reach the age of 60
- design reinforcement and detail bars to be of sensible weight, the fewer bars to be fixed the better, and having a minimum number

of bar types (maximise repetition). This will also speed site progress
- avoid the use of heavy concrete blockwork. 200 mm thick solid concrete blocks are too heavy to consider in construction, yet they are specified on some projects. The result is very slow progress. The solution is to lay 100 mm blocks on the flat
- cement is now supplied in 25 kg sacks — for years the industry struggled (and wasted) as it used 50 kg sacks
- store items with future lifting in mind
- design plant and other fixed items with future lifting in mind — the provision of adequate and correctly placed lifting beams, for example
- consider the access needed for correct lifting to take place
- design for prepack, heavier lifts, pre-assembly, and lift heavier loads using appropriate mechanical assistance. House builders are prefabricating block and brick panels, and are using heavy forklift trucks to position them:
  - bespoke reinforcement, designed on a repetitive basis, can be fabricated off-site (preferably using welding technique) and lifted into position
  - projects are specifically targeting wastage by supplying correct amounts of their sub-contractors at the required time.

Lifting or working in an unsuitable manner

- Steelfixers bending to fix reinforcement at ground level, in slabs and beams, suffer health problems due to repetitive back strain. Designs should consider pre-assembly of beams at ground level using waist-high saddles or fixing tables to work on, then hoisting the completed beam into position with cranes. In a similar way, slabs could be prefabricated at waist level and hoisted into position. Designs need to be of sufficient size to allow this and construction will be quicker as well as safer.
- Trenches in the ground need to be of a width which enables a person to work safely and comfortably in them. The minimum trench width, as required by the design, is often too narrow for this to happen.
- Items in a structure which need later maintenance should be designed with safe and comfortable access in mind. Better maintenance will result as well as increased safety.

- Items in a structure which will need replacement need to consider the replacement process in the design.
- Design to allow adequate working space — a better, safer job will result.

Deafness and ear problems

- Plant is now designed to be much quieter.
- Ear defenders are provided and used on a very regular basis.
- Designers should avoid variations which are noisy to execute.

There is a strong argument that the application of suitable safety measures will not only give compliance with the regulations, but also:

- make work safer
- enable operatives to work more efficiently
- increase the usage of attendant plant
- speed construction progress
- reduce costs.

A practical design, which considers buildability in parallel with the relevant design codes, is necessary if this is to be achieved.

Chapter 3

The business requirement

Successful construction management

Experience of the construction industry shows that each project, while perhaps similar to many others, will be unique in itself and need to be given its own required and separate consideration. Some of us have undoubtedly relished the prospect of doing work similar to what we have done previously — and regretted later that practice on one job may not work so well as on another. Some jobs make money, others lose money.

So, by and large, we take each project on its merits and set up a suitable project team which will see the project to completion. While the project itself will be unique, it will use any of a variety of management considerations which are common to all projects, whether these are in the industry or not. These considerations include the following.

(*a*) *Competency* — no one would wish to appoint a person or a team deemed incompetent. And we mean competent relevant to the project envisaged. The CDM Regulations demand such competency.

(*b*) *Adequacy* — with the best will in the world, we will fail in our task if we fail to provide adequate resources to carry out a project. The CDM Regulations demand that we provide adequate resources.

(*c*) *Time* — time to do a job properly. Time to get it right first time. Crash programmes often give less then desirable results. The CDM Regulations note that time is a resource and must be adequate.

(*d*) *Teamwork* — perhaps the best way to consider teamwork is to ask who would prefer to work without it? Teamwork gives us the opportunity to use the skills of others and not have to rely on ours alone. It encourages others to take ownership of the task. The job becomes theirs rather than ours. And people will strive harder to ensure that their job is successful than they will to make our job successful. The team will work better as it becomes more adept. A key element of management training is the development of teams, individuals and oneself.

(*e*) *Effective communications* — if the project is to be successful, it is essential that the end product complies fully with that which the client requested initially. This requires a clear message to be passed down the construction chain. That those producing the project produce that which was intended.

(*f*) *Fairness* — do you like being treated unfairly? Of course you do not! And nor does anyone else. Can we expect others to perform well for us if we pay them late or too little, or if we impose unfair contract conditions on them? Doubtful! The Latham Report (1994) stresses the need to deal fairly with others and to use forms of contract which are based on fairness.

(*g*) *Honesty, integrity and truthfulness* — these characteristics, and other similar ones, define the sort of people we enjoy working with. Characteristics we take to refer to us also.

If we can select a project team which includes these skills, then we really do have an opportunity to perform. A capability that will help us to get it 'right first time all down the line'.

The desired team is unlikely to be the cheapest. At the same time, however, you often get the high performer without it costing too much extra. If we can get a top performance team, should we be denied the opportunity of using such a team because the competition is a bit cheaper? The National Audit Office recently recommended that Government cease awarding contracts on the basis of lowest price — a policy said to be costing the economy £2 billion per annum.

Whatever title we give it, successful project management starts by putting together the best team that you can. One that has the relevant practical skills necessary to understand and carry out all parts of the project, and ensuring that the various parts of the team, and the individuals within the teams, work together properly. We will then have a good chance of getting it 'right first time'.

Allocation of risk

Latham (1994) recommends that clients carry out an internal risk assessment on a proposed project and then decide how the risk inherent should be allocated between the parties. Various types of contract are considered from a risk point of view and the advantages and disadvantages of various contract strategies are considered. The following paragraphs 3.7 to 3.12, 3.16 and 3.17, together with Tables 2 and 3 and Box 3, are taken directly from the Latham Report.

Contract strategy and risk assessment

3.7 Once a client is satisfied about real need and feasibility within overall budgetary constraints, the instinctive reaction is to retain a consultant to design the project — the 'ring up an architect/engineer' syndrome. That takes a crucial step too quickly, and closes off potential procurement options. The next step should be the use of internal risk assessment to devise a contract strategy. The client should decide how much risk to accept. No construction project is risk free. Risk can be managed, minimised, shared, transferred or accepted. It cannot be ignored. The client who wishes to accept little or no risk should take different routes for procuring advice from the client who places importance on detailed, hands-on control.

3.8 The basic decision on the procurement route should precede the preparation of the outline (project) brief, since it necessarily affects who shall assist with the design brief as well. That choice of route must be determined by the nature of the project and the clients' wishes over acceptance of risk. Such decisions are difficult. Inexperienced clients need advice. There are a number of publications which can assist.

3.9 Tables 2 and 3 summarise the distribution of risk under the respective Standard Forms of Contract, and how the client may broadly assess it in advance

3.10 Guidance on risk assessment for clients is crucial. It determines contract strategy. A range of procurement and contractual routes is emerging to meet clients' wishes. They are set out in box 3.

The brief

3.11 Once a prospective client has decided that a project should proceed in principle, and roughly how much risk and direct

Table 2. Contract options (taken from the Latham Report (1994))

	Risk	
	Client	Contractor
Management		
Prime cost:		
Percentage fee		
Fixed fee		
Approximate quantities:		
Remeasured		
Lump sum:		
Fluctuations		
Fixed price		
Design and build		
Package deal		

Key:

Fundamental risks

Pure and particular risks

Speculative risks

1. Fundamental risks: war damage, nuclear pollution, supersonic bangs
2. Pure risks: fire damage, storm
3. Particular risks: collapse, subsidence, vibration, removal of support
4. Speculative risks: ground conditions, inflation, weather, shortages and taxes

involvement to accept, the project and design briefs can be prepared. The client who knows exactly what is required can instruct the intended provider. That may involve either appointing a project manager, or a client's representative, to liaise with the designers, or a lead designer, or a contractor for direct design and build procurement. Even the best clients are likely to benefit from some advice on alternative methods of achieving

Table 3. Summary of advantages and disadvantages of contract strategies (CUP Note. This table is for guidance only. Generally the appropriateness of the contract is not as clear-cut as indicated. The project manager must advise the project sponsor on this) (table taken from the Latham Report (1994))

Project objectives		Appropriateness of contract strategy in meeting project objectives				
Parameter	Objectives	Traditional	Construction management	Management contracting	Design and manage	Design and build
Timing	Early completion	□	■	■	■	■
Cost	Price certainty before construction starts	■	□	□	□	■
Quality	Prestige level in design and construction	■	■	■	□	□
Variations	Avoid prohibitive costs of change	■	■	■	■	□
Complexity	Technically advanced or highly complex building	□	■	■	□	□
Responsibility	Single contractual link for project execution	□	□	□	■	■
Professional responsibility	Need for design team to report to sponsor	■	■	■	□	□
Risk avoidance	Desire to transfer complete risk	□	□	□	□	■
Damage recovery	Ability to recover costs direct from the contractor	■	□	■	■	■
Buildability	Contractor input to economic construction to benefit the department	□	■	■	■	□

■ Appropiate □ Not appropriate

Box 3 (taken from the Latham Report (1994))

Standard construction
The client and its advisers decide that the end product can be achieved through a pre-determined construction route, probably involving a limited range of standardised processes and components. This is best served by a design and build contract, with single point responsibility and the transfer of risk by the client. The contractor will be responsible for delivering the entire package. The amount of design provided by the client to the contractor will vary. It may involve the novation of some consultants to the contractor. The client retains an employer's agent to liaise with the contractor.

'Traditional' construction
These projects involve well-used and normal techniques of design and construction, but reflect specific wishes of the client. Most work is currently done on this basis, involving Standard Forms such as JCT 80 or ICE 5th/6th. This is the route with which the industry is most familiar. But it is also where many of the problems emerge through lack of co-ordination between design and construction. Problems can however be minimised or avoided by effective pre-planning of the design process, and efficient administration of the project by the client's representative, whatever title or designation that representative is given. An alternative is to use a combination of routes through design and manage approaches. The consultants, working for a lead manager, design and cost the scheme within their normal professional roles. They can then be paid their fees and discharged, after certifying that their responsibilities for the design are complete. The implementation of the design is then passed to a management contractor who is paid a fee to mobilise a series of works contractors to build the project. The client retains an employer's agent to liaise with the contractor.

Innovative construction
The client commissions a project which involves a high degree of innovation, and many new design details. The client wants hands-on involvement and seeks strong management to produce

> the intended result. The best route here is construction management. It is a demanding procurement system, requiring firm leadership and teamwork throughout. There is no intrinsic reason why it should be limited to large or exceptionally prestigious schemes. It is also favoured by specialist/trade contractors because of its separate trade contract system. Some main contractors now offer this service, as do specialist construction management companies.

their aim, which may produce better value for money. Most clients require detailed advice. Getting the design brief right is crucial to the effective delivery of the project. The lead adviser must be given time to assist the client with the preparation of that brief. This should be an iterative process. Stages A and B of the Royal Institute of British Architects (RIBA) 'plan of work' are described as 'inception' and 'feasibility'. The client will already have taken some basic procurement decisions on these matters. But whether or not the scheme is designer-led, the client must allow time and space for its wishes as expressed in the project brief to be further tested for the purpose of the design brief against questions of feasibility. Time may also be needed for other advisers — including the essential mechanical and engineering design/construction input — to be called in at these early stages. Commercial pressures from the client may require the detailed designs to be prepared sequentially. But clients will benefit by allowing enough time for a good brief to be devised in order to avoid subsequent delays and cost overruns in the project.

3.12 There should be a distinction between the project brief — the basic objectives of the client — and the design brief which comprises the client's specific requirements. The CIC [Construction Industry Council], in its final report, recommends a comprehensive standard questionnaire and checklist to assist the client in the briefing process. But RIBA believes that a 'check list of standard subjects will necessarily be too general and rudimentary to be of much relevance' (RIBA evidence, April 1994). The CIEC's [Construction Industry Employers Council] final report also favours an 'authoritative guide' for clients to help 'produce a comprehensive, agreed design brief', and points out that a good deal of work has already been done in this regard by various organisations.

Procurement for large process/power plant projects

3.16 Large engineering projects in the process plant sector of construction require detailed design and planning before commitment. The contract strategy may involve performance specifications being issued by the client, evidence being required that contractors have successful experience of a similar project, and possibly a condition that the successful tenderer should operate the plant or product after completion. The managing contractor is likely to offer full Engineer-Procure-Construct (EPC) capabilities. The European Construction Institute, which gave evidence to the Review, both directly and through the CIPS [Chartered Institute of Purchase and Supply], stressed:

1. Economies at the design phase will be self-defeating. Designers should not be selected on the basis of price. 10–15% of the total cost of a high technology project should be spent in this phase.
2. Where prototype equipment is involved, it must be identified in the initial stage of the project. Detailed programmes for research, design, testing and manufacture must be produced and monitored by the client. This should all be available before work starts on site.
3. Failure of material flow to the site or design changes can lead to unmanageable situations. Designs should be frozen and fully developed before the manufacture and site construction begins.
4. A 'key date' procedure for discussing cash flow and resources at a very high level is vital every three months or so, as well as normal project progress meetings. Payments should be made properly and on time, provided that milestones have been achieved. The project manager should also be aware of whether the main contractor is paying the sub-contractors on time, so as to prevent problems on site.

3.17 Effective partnering between client and contractor with teamwork and a 'win-win' approach helped to bring the Sizewell nuclear power station to completion on time and within budget. But that client believes that there is still scope for further improvements in productivity and cost reduction, especially if design and construction teams could be kept together.

While each project must be considered on its own merits, the guidance offered reflects a common sense approach. The prime need is to get a project 'right first time'. The process starts at project conception. Guidance on the necessary inclusions in the contract itself is also given by Latham.

The contract choice for clients

The Latham Report (1994) suggests that the most effective form of contract in modern conditions should include the following.

1. A specific duty for all parties to deal fairly with each other, and with their sub-contractors, specialists and suppliers, in an atmosphere of mutual cooperation.
2. Firm duties of teamwork, with shared financial motivation to pursue those objectives. These should involve a general presumption to achieve 'win-win' solutions to problems which may arise during the course of the project.
3. A wholly interrelated package of documents which clearly defines the roles and duties of all involved, and which is suitable for all types of project and for any procurement route.
4. Easily comprehensible language and with guidance notes attached.
5. Separation of the roles of contract administrator, project or lead manager and adjudicator. The project or lead manager should be clearly defined as the client's representative.
6. A choice of allocation of risks, to be decided as appropriate to each project but then allocated to the party best able to manage, estimate and carry the risk.
7. Taking all reasonable steps to avoid changes to pre-planned works information. But, where variations do occur, they should be priced in advance, with provision for independent adjudication if agreement cannot be reached.
8. Express provision for assessing interim payment by methods other than monthly valuation, i.e. milestones, activity schedules or payment schedules. Such arrangements must also be reflected in the related sub-contract documentation. The eventual aim should be to phase out the traditional system of monthly measurement or remeasurement but meanwhile provision should still be made for it.

9. Clearly setting out the period within which interim payments must be made to all participants in the process, failing which they will have an automatic right to compensation, involving payment of interest at a sufficiently heavy rate to deter slow payment.
10. Providing for secure trust fund routes of payment.
11. While taking all possible steps to avoid conflict on site, providing for speedy dispute resolution if any conflict arises, by a pre-determined impartial adjudicator/referee/expert.
12. Providing for incentives for exceptional performance.
13. Making provision where appropriate for advance mobilisations payments (if necessary, bonded) to contractors and sub-contractors, including in respect of off-site prefabricated materials provided by part of the construction team.

The New Engineering Contract (NEC) is stated to cover virtually all these assumptions of best practice. Feedback on the NEC by those using it is largely positive. It is felt to encourage project teams to work together more closely, helps resolve problems more quickly and avoids the need to involve lawyers. Accounts are felt to be resolved more quickly.

Lawyers have expressed a reluctance to use NEC whose plain English style has yet to be tested in the courts. Must we always be slaves to the past? Guidance from HM Treasury states the following.

(*a*) The contract requirement sets out what the consultant or contractor is required to do under the contract. The brief or specification forms a key part of the requirements and should be output based. Where practicable, the requirements should include targets and milestones that are achievable and measurable.

(*b*) The use of standard forms of contract helps to reduce both tendering and contract administration costs. Bespoke or amended forms require departments and tenderers to seek additional and frequently costly legal advice and this increases the risk of disputes arising from unfamiliar terms. Therefore, to avoid unnecessary additional costs, such forms should only be used where they are considered essential rather than simply desirable and where they demonstrably provide greater value for money. Any amendments should only be made after receiving technical and legal advice.

(*c*) Ideally, suites of contracts and standard unamended contract forms from recognised bodies should be used where they are available.

Designs and drawings

The drawings define, to a large degree, how a project will be put together. If this is easy, the work will be cheap, if difficult, expensive. Contractors seek involvement in the design process in order to simplify the process. They bring the benefit of their construction experience to the designer. The logical intention must be to achieve a safe and simple construction process. The designer, perhaps lacking in such experience, could well fail at this stage.

Simplification will lead not only to reduced costs but to an easier-to-achieve programme, easier resourcing and better levels of safety. The work itself will be easier to manage. Difficult construction causes the reverse to apply. Cost increases and programmes are slowed when large number of small items are used (blocks, for example), rather than small numbers of large items (precast panels, for example). Complex joints and fixing arrangements increase cost and delays. Construction on skew angles rather than at 90° can prove very expensive and show little benefit. Tolerances should fall within acceptable limits. They must be sufficiently tight to ensure efficient erection, yet sufficiently high to allow sensible construction. A tolerance which is too tight can make work impossible to carry out while too slack a tolerance can lead to an unacceptably large loss of quality. Tolerance varies between materials and processes. In-situ concrete requires a higher tolerance than precast work. Good buildability requires compatibility between the differing relevant tolerances.

Construction work is made more difficult, slower and more costly, when a large number of different parts or materials are used. Wastage is likely to increase, the workforce will need a wider range of skills. We may perhaps need a wider range of plant.

Reinforcement is easier to fix if we have fewer bars of larger diameter and consideration is given to the weight of the individual bars.

Handling, getting the right pieces to the right place at the right time, is made more difficult as the variety of components gets greater. We might even have to consider a correct sequence of loading and unloading of vehicles delivering to site in extreme cases, especially if parts are erected immediately upon delivery. Precast construction in a busy urban area could be an example.

Once variety has been minimised, we need to increase the frequency of use of items — be more repetitive. The workforce will work more efficiently by virtue of the repetitive process, there will be fewer design options to consider at the interfaces, and organisation will be simpler.

Structural steel connections, precast concrete items and fixing bolts are examples of areas of work where standardisation and repetitiveness can reduce errors, simplify the work itself and increase productivity.

Dimensions should be rationalised as far as possible. To have, for example:

- columns, beams and slabs of a standard size, or at least to a repetitive pattern of sizes
- columns at centres which suit the infill between them — if, for example, a 1200 mm board is to be used, consider 300, 600 or 1200 modules to fill the gap
- standard ceiling heights with proper allowance for false ceilings and future use and access
- rationalised lengths — of reinforcement bars, blocks, bricks and windows, for example.

Labour-intensive work — concrete, brickwork, blockwork are examples of this — tends to be slow. As following work often has to be delayed until the structure is watertight, then following trades are all delayed. It should be avoided where possible.

Adequate and simple fixings should be provided for all items. An expensive item (a façade, for example) should not have to be replaced due the failure of relatively cheap fixings. Insure for success in the design.

Overhangs on buildings, at the roof for example, or over windows, can make later maintenance difficult and expensive.

We should incorporate total buildability as a key factor from the onset of a project and include this in the feasibility study, paying particular attention to time and cost. Research, or seeking opinion of those who have practical experience of the various issues, all helps. Consider everything and get it right first time.

Basic construction, although often standard, needs to consider buildability as a focus for change and improvement of the process if improvements are to be made.

Innovative work needs detailed consideration of its buildability, simply because it is innovative. We hear many instances of problems occurring on such projects simply because they have been designed to be innovative and no thought given to buildability. This can prove expensive.

It is important that the design considers the conditions on the site itself which will affect the construction process.

(*a*) Work below ground can pose particular problems. Are extensive temporary works required? Is the site contaminated or wet, or are adjacent buildings going to be affected? Lack of consideration at the design stage can create construction risks later. The intention is to avoid such risks occurring and to simplify the construction requirement.

(*b*) The provision of the services in a structure needs to be facilitated in a way which does not affect either the construction itself or the future needs of the client (of access, for example).

(*c*) Access to a project is a key factor. Cramped sites affect site movement, plant and operations generally are restricted, the siting of effective scaffold can be a problem, storage is likely to be a continuing issue, neighbours in adjacent structures require consideration.

(*d*) The requirement for ease of maintenance and replacement of parts in the finished structure requires careful consideration. Plant rooms need to be adequate in size, located and fitted out in a manner which will allow easy maintenance, change over of equipment and so on. A retailer, with lifts which cater for objects up to 2 m high, elected to use new shelving which is 2·4 m high. Shelving has to be partially dismantled to fit into the lifts. When it is erected to the full 2·4 m height it is unsafe to use the existing shelf filling arrangements on them. Window cleaning, staircases designed for speed of erection (in steel or precast concrete) and the possible requirement of additional space to enable increased future services, can all affect project viability and life-time costs. Structures are demolished when they are incapable of being modified to suit future needs. This can be wasteful!

Buildability can be adversely affected by splitting the design from construction, some designers being unfamiliar with what is practical on site. It is common sense to utilise the experience of a contractor who knows how to do the job.

Designers need to have the ability to prepare the design. Is the job too big or too small? Is the timing right? Are adequate resources available?

Designs require adequate site investigations to support them. Too many errors occur due to the provision of inadequate or incorrect data.

Structures should be designed, fitted out and have layouts which maximise performance in use, can preferably be adapted for a change of use (they are not *too* specialist) and can be sold for a good price when requirements change.

Ease of maintenance and replacement are important buildability considerations. Lifelong costs may exceed those of the initial construction. Durability is an important consideration, not just in terms of how long an item will last, but how easy it is to damage. Efficient durability involves the selection of items which wear out progressively during the life of a structure. You do not need to spend on items which last beyond that life. Items of poor quality tend to increase maintenance costs and have to be replaced more often. We may benefit by spending a little more and getting better quality in the first place.

Maintenance should ideally be carried out by local resources, especially in developing countries, and designers need to consider simplicity and ease of maintenance. Complicated maintenance, perhaps using specialist plant and highly skilled labour, could be a problem in such situations.

Services are likely to need updating during the life of a structure. Allowance for this, including the provision of extra space, should be included in the design. The use of both false floors and ceilings should be considered.

Some fitments will require replacement progressively through the life of a structure. Provision should be made for new items to be fitted into old surrounds without the replacement of anything else. New radiators should fit to existing pipework. New doors and windows should fit the old recesses. Demountable partitions, wall panels, floors and ceilings should be designed to accommodate alterations in factors such as cladding or floor loadings.

When specifying items, the saving of money by specifying cheap items of lower quality can lead to increased maintenance costs later. On the other hand, the provision of unnecessary quality can prove expensive, it may be damaged by cheaper items or even look out of place.

The drawings convey the construction requirement to the tradesmen and operatives who build the project. In common with the client and the contractor, they are seeking construction details which are easy to understand, quick to implement and safe to provide. They can then set out to do a good job and earn reasonable payment in the process. The opposite case, using impractical details, is to be avoided.

To avoid errors, drawings should 'stand alone' — be complete in themselves. This will avoid the errors which occur due to the omission of detail which appears on other drawings. When drawings are read on site to understand the work, the user will be much more effective with a single sheet than if he is forced to struggle with several.

Bill of Quantities

A Bill of Quantities is essential if we are to accurately define the full workscope of a project. While it is possible to assess the likely cost of a project on the basis of the price per cubic metre of the concrete required or the price per square metre of floor area provided, any cost projection based on these will be approximate.

Building works, where many items of the work (painting, plastering, decorating and tiling, for example) are repetitive and attract similar prices on every contract, lend themselves more easily to such approximate methods than civil engineering work, where each project has its differences to other similar projects.

An accurate definition of the workscope will enable later management decisions to be taken on the basis of total, rather than partial information. A definition needs to be complete, accurate in its parts, and explained in a manner which is readily understood by those concerned. You need a standard method of measurement. A standard method, just like a checklist or crib sheet, enables anyone carrying out relevant work to ensure that everything is covered. They do not have to reinvent the wheel every time they do the task. Standardisation helps everyone involved in the process to understand what the other party is proposing. It enables easier understanding.

The *Civil Engineering Standard Method of Measurement* (CESMM) (ICE, 1991) is a standard method. Its object is to set out the procedure to be followed when preparing a relevant Bill and how to express and measure the quantities of work. The objectives of the Bill of Quantities are as follows.

(*a*) To provide such information of the quantities of work as to enable tenders to be prepared efficiently and accurately. In a competitive commercial situation, it is foolhardy to approximate.

(*b*) When a contract has been entered into, to provide for use of the priced Bill of Quantities in the valuation of work executed. It enables the accurate ordering of materials and sub-contract services and of correct and verifiable stage payments to the parties concerned.

In addition, an accurate Bill of Quantities will:

- enable the content of drawings to be checked — if you cannot measure it, you cannot build it!
- enable clients to accurately define workscope and price at an early stage
- enable allowance to be made for provisional items of work, not shown on the drawings, to be catered for (unsuitable material, rock, water ingress, for example)

CESMM defines:

- how work is to be divided into separate items in the Bill of Quantities — standard lists of items
- the information to be given in item descriptions — standard descriptions of items enabling fair billing and fair pricing to take place
- the units in which the quantities against each item are to be expressed (no., m^3, m^2, m and so on) — this enables standard pricing methods to be used
- how the work is to be measured for the purpose of calculating quantities.

CESMM developed from the *Standard Method of Measurement of Civil Engineering Quantities* (ICE, 1953), which was used by the ICE until 1976. It was felt that a Bill of Quantities, which was, in essence, no more than a price list of permanent works, no longer adequately reflected the many variables in the cost of civil engineering construction that have resulted from developments in constructional techniques and methods. The improvements sought were to:

- standardise the layout and contents of Bill of Quantities prepared according to the standard method of measurement
- provide a systematic structure of Bill items leading to more uniform itemisation and description
- review the subdivision of work into items so that a more sensitive and balanced description of the work in a contract is provided.
- take account of new techniques in civil engineering construction and management, their influence on the work itself and on the administration of contracts.

A standard method, prepared using independent stated guidelines and understood by all parties is clearly better than an 'ad hoc' presentation. The introduction to *The CESMM 3 handbook* by Dr Martin Barnes (1992) states:

> Financial control means control of money changing hands. Since money almost always changes hands in the opposite direction from that in which goods or services are supplied, it can be considered as the control of who provides what and at what price. This thought establishes a priced Bill of Quantities as the central vehicle for the financial control of a civil engineering contract. The Bill of Quantities is the agreed statement of the prices which will be paid for work done by the contractor for the employer

and it shares with the drawings and the specification the responsibility for defining what has been agreed shall be done.

Control is usually based on a forecast. The difficulty of controlling something is proportional to the difficulty of predicting its behaviour. The points, finer and coarser, of the financial control of civil engineering contracts revolve around the difficulty the employer has in forecasting and defining to a contractor precisely and immutably what he is required to do, and the difficulty the contractor has in forecasting precisely what the work will cost. To achieve effective control it is necessary to limit these difficulties as much as possible within reasonable limits of practicality. This means using as much precision as possible in defining the work to the contractor and in enabling him to forecast its cost as precisely as possible. These are the essential functions of Bills of Quantities. It is the essential function of a method of measurement to define how Bills of Quantities should be complied so that they serve these two essential functions.

It is clear from this consideration that a Bill of Quantities works best if it is a model in words and numbers of the work in a contract. Such a model could be large, intricately detailed and reproducing the workings of the real thing in an exact representation. Alternatively, it could be as simple as possible while still reproducing accurately those aspects of the behaviour of the original which are relevant to the purposes for which the model is constructed.

The first purpose of a Bill of Quantities is to facilitate the estimating of the cost of work by a contractor when tendering. Considered as a model, it should therefore comprise a list of carefully described parameters on which the cost of the work to be done can be expected to depend. Clearly these parameters should include the quantities of the work to be done in the course of the main construction operations. There is no point in listing those parameters whose influence on the total cost of the work is so small as to be masked by uncertainty in the forecasting of the cost of the major operations.

Other points of general application emerge from this principle of cost-significant parameters. The separation of design from construction in civil engineering contracts and the appointment of contractors on the basis of the lowest tender are the two features of the system which make it essential for a good set of parameters to be passed to contractors for pricing, and for a good set of priced parameters to be passed back to designers and employers. Only then can they design and plan with the benefit of realistic knowledge of how their decisions will affect construction costs. The less contractual pressures cause distortion of the form of the prices exchanged from the form of actual construction costs the better this object is served. It is very much in the interests of employers of the civil engineering industry, whether they are habitually or only occasionally in that role, that the distortion of actual cost parameters should be minimised in priced Bills of Quantities.

An employer's most important decision is whether to proceed to construction or not. This decision, if it is not to be taken wrongly, must be based on accurate forecast of contract price. Only if a designer has a means of predicting likely construction cost can such a forecast be achieved. The absence of cost parameters which are sensitive to methods and timing of construction has probably caused as much waste of capital as any other characteristic of the civil engineering industry. It has sustained dependence on the view that quantity is the only cost-significant parameter long after the era when it had some veracity. Generations of contractors, facing drawings first when estimating, have found themselves marvelling at the construction complexity of some concrete shape which has apparently been designed with the object of carrying loads using the minimum volume of concrete. That it has required unnecessary expense in constructing the formwork, in bending and placing the reinforcement and in supporting the member until the concrete is cured often appears to have been ignored.

A major aspect of financial control in civil engineering contracts is the control of the prices paid for work which has been varied. Varying works means varying what the contractor will be required to do, not varying what has already been actually done. Having once been built, work is seldom varied by demolition and reconstruction; the difficulty of pricing variations arises because what gets built is not what the contractor originally plans to build. If the work actually built were that shown on the original drawings and measured in the original Bill of Quantities, the prices given would have to cover all the intricate combinations of costs which produce the total cost the contractor will actually experience. This would include every hour of every man's paid time — his good days, his bad days and the days when what he does is totally unforeseen. It would include every tonne or cubic metre of material and the unknown number of bricks which get trodden into the ground. It would include every hour of use of every piece of plant and the weeks when the least popular bulldozer is parked in a far corner of the site with a track roller missing. The original estimate of the total cost of this varied and unpredictable series of activities could reasonably only be based on an attempt to foresee the level of resources required to finish the job, with many little overestimates balancing many little underestimates. Changes to the work from that originally planned may produce changes to total cost which are unrelated to changed quantities of work. They are less likely to produce changes in cost which are close to the changed valuation if value is taken to be purely proportional to quantity of the finished work. Where there are many variations to the work the act of faith embodied in the original estimate and tender can be completely undermined. The contractor may find himself living from day to day doing work the costs of which have no relationship to the pattern originally assumed.

That cost is difficult to predict must not be allowed to obscure the fact that financial control depends on prediction. If the content of the work cannot be predicted the conduct of the work cannot be planned. If work cannot be planned its cost can only be recorded, not controlled. It must also be accepted that valuation of variations using only unit prices in bills of quantities is an unrealistic exercise for most work and does little to restore the heavily varied project to a climate of effective financial control. Only for the few items of work whose costs are dominated by the cost of a freely available material is the quantity of work a realistic cost parameter. It follows that employers are well served by the civil engineering industry only if contractors are able to plan work effectively; to select and mobilise the plant and labour teams most appropriate to the scale and nature of the expected work and to apply experience and ingenuity to the choice of the most appropriate methods of construction and use of temporary works.

That this type of planning is often invalidated by variations and delays has blunted the incentive of contractors to plan in the interests of economy and profit. The use of over-simplified and unrealistic parameters for pricing variations has led to effort being applied to the pursuit of payment instead of to the pursuit of construction efficiency. In a climate of uncertainty brainpower may be better applied to maximising payment than to minimising cost.

Mitigation of this problem lies in using better parameters of cost as the basis of prices in Bills of Quantities. It would be ideal if the items in a Bill were a set of parameters of total project cost which the contractor had priced by forecasting the cost of each and then adding a uniform margin to allow for profit. Then, if parameters such as quantity of work or a length of time were to change, the application of the new parameters to each of the prices would produce a new total price bearing the same relationship to the original estimated price as the new total cost bore to the original estimated cost. The employer would then pay for variations at prices which were clearly related to tender prices and the derivation of the adjusted price would be wholly systematic and uncontentious. This ideal is unobtainable, but it is brought closer as Bills of Quantities are built up from increasingly realistic parameters of actual construction costs.

From the cash flow point of view there are also advantages in sticking to the principle of cost parameters. The closer the relationship between the pattern of the prices in a Bill of Quantities and the pattern of the construction costs, the closer the amount paid by the employer to the contractor each month is to the amount paid by the contractor each month to his suppliers and sub-contractors. The contractor's cash balance position is stabilised, only accumulating profit or loss when his operations are costing less or more than was estimated.

Since most of the contractor's turnover is that of materials suppliers and sub-contractors with little added value, stability and predictability of cash flow has an importance often not appreciated by employers and engineers. Contractors are in business to achieve a return on their resources of management and working capital — a return which is seldom related closely to profit on turnover. Predictability of the amount of working capital required is a function of prompt and cost-related payment from the employer, another benefit of using pricing parameters closely related to parameters of construction cost.

An employer's interest is best served by a contractor who is able to base an accurate estimate on a reliable plan for constructing a clearly defined project, and who is able to carry out the work with a continuing incentive to build efficiently and economically despite the assaults of those unforeseen circumstances which characterise civil engineering work. Confidence in being paid fully, promptly and fairly will lead to the prosperity of efficient contractors and to the demise of those whose success depends more on the vigour with which they pursue doubtful claims.

As Louis XIV's department of works was recommended in 1683, as a result of what may have been the first government enquiry into the financial control of civil engineering contracts: 'In the name of God: re-establish good faith, give the quantities of the work and do not refuse a reasonable extra payment to the contractor who will fulfil his obligations.'

Specifications

The specification for a project is written to define the performance requirement. Performance may be described in terms of:

- strength — of a material
- standard — usually a British Standard
- type — the use of engineering bricks
- design — the reinstatement of footpaths giving the required minimum thicknesses of materials, for example
- tolerances — how accurate to line and level the required work must be, and so on.

Specifications cover aspects of:

- general requirements, e.g.:
  - entry onto site
  - fencing
  - surveys
  - contamination
  - engineers' requirements
- materials: the performance requirement of every material required for the project will be specified — strength, appearance, where to be used are all relevant
- different aspects of work — excavation, concreting, pipelaying, brickwork and so on. The performance requirement for each aspect is specified.

Some specifications will concentrate on performance, leaving the contractor to decide his method of working. Such a specification gives a wide range of options for carrying out the work. The contractor can select his methods and materials based on 'best value' at the particular time and place. Materials will be chosen on the basis of availability and price. Working methods will be those which best suit the contractor. The civil engineering specification for the water industry is an example of this.

Other specifications will specify product as well as performance. The type of brick or cladding or other material might be specified. It is more specific. It would apply to a project whose architect desires to use certain products to produce a particular effect. Issues such as availability, cost, compatibility between items, indeed suitability in general, need to be carefully considered when drawing up such specifications. Lack of choice sometimes creates expensive problems.

Some clients, responsible for continuous construction work, will have a general specification, which covers most performance requirements, and a specific specification, which covers those requirements which are specific to a particular project.

Whichever specification is used, it is clear that project costs are very much influenced by the specified requirement. Costs will increase if:

- expensive materials are specified
- materials are difficult to obtain
- fine tolerances on workmanship are applied.

Costs will decrease if the reverse is true. Architects, designers and those drawing up specifications need to be aware of the cost implication.

Fine tolerances, on brickwork, concrete finishes and formwork for example, have a major effect on programme and cost in comparison to more reasonable, and perfectly acceptable, wider tolerances. Most contractors have avid experience of the difficulties created by such specifications.

Discussing issues affecting buildability in a large organisation, a senior design engineer repeatedly commented that 'We have specific and generic specifications on each contract. It is not to us to tell the contractor how to perform'. This is correct insofar as it goes. If, however, we are to seriously consider and implement the recommendations of Latham (1994), Egan (1998) and the regulations relating to safety and the cost of our projects, then wider considerations need to be made. Best practice should be a continuous goal which is never reached. The engineer in question admitted to having no site experience. Nor did he question the content of the specification.

The experienced practitioner will be fully aware of the full implications of a specification which is not fully suited to the project requirement.

From client to end user — and back again

Client

(*a*) Egan (1998) says that in the best companies the client drives everything. These companies provide exactly what the end customer needs, when the customer needs it and at a price that reflects the product's value to the customer (see chapter 17 'A focus on the customer' in the Egan Report). This can only be achieved if the client states the requirement *exactly*.

Latham (1994) recommends that a client first satisfies himself about the real need for a project and its feasibility within overall budgetary requirements. Assuming feasibility, a risk assessment should be undertaken and the client should then decide how much risk to accept. This will define the contract strategy to be adopted. Only when feasibility and strategy are decided should a consultant be retained (see sections 3.4–3.6 'Project strategy and need assessment').

(*b*) In terms of good communication the client now needs to define *exactly* what he wants, when he wants it, and where it is to be. This optimises the client's chances of getting *exactly* what he wants. Too many projects fail because the client is initially unsure of what he wants. The progressive finalising of ideas as the project proceeds, whether it be the planning, design, construction or later phase costs time and money. And little positive benefit is derived from the late thinking.

While others, the architect, designers and the contractors involved will have to ensure their own communications, the client ultimately pays most of the bill, if not all of it. He, therefore, has a vested interested in specifications, drawings, contracts and details of all types which are easily understood and executed. He should ensure that this is the case.

Communication will be two ways if it is to be fully effective. We need feedback from all parts of the supply chain to tell us how things are progressing. What is going right and, sometimes more importantly, what is going wrong. It we quickly find out what is going wrong, we can take corrective action quickly — and minimise the size and cost of any defect.

(*c*) The client has a duty under the CDM Regulations to provide information on the site. Incorrect information — of ground conditions, contamination or the state of an existing structure, for example — can ultimately lead to large increases in cost. Accurate data can save money as well as satisfy a legal requirement.

Good communication will ensure that *full* and correct information is provided to those who might need it — the planning supervisor, the designers, the principal contractors and reduce, perhaps eliminate, any claims which arise.

(*d*) Who should the client select as his architect or designer? Competition and fee schedules imply that costs will be similar between various practices. Their cost will also be a relatively small percentage of the overall cost. Is there any benefit in going for the lowest price? Probably very little. The Audit Commission has recently advised the Government that opting for lowest cost actually *costs* money?

There is a compelling logic, however, in seeking a partner who has the relevant skills for the job and can make those skills available in adequate numbers at the appropriate time. Such a partner is likely to be able to do the job in a satisfactory matter. If the parties have worked together previously, mutual confidence and respect are likely to occur from the onset. The client can benefit from the findings of Latham and Egan as well as the CDM Regulations.

Guidance given to government departments by HM Treasury (1997) on the appointment of consultants and contractors states:

> Departments should make a public policy commitment to give a client lead to the industry by:
>
> - openly finding out which designers, constructors or specialist suppliers are the best;
> - tendering with the aim of getting those who offer the best service;
> - working with their people as a team, not opponents;
> - making no compromises with people or suppliers who are uncooperative or adversarial.

The guidance gives key points for senior management to note.

(*a*) UK Government policy is that all procurement should be on the basis of value for money and not lowest price alone. Consultants and contractors therefore should be appointed on this basis.

(*b*) Robust mechanisms specific to each project should be developed to evaluate the quality and price (whole life cost) components of each bid in a fair, transparent and accountable manner.
(*c*) The selection and award processes are separate and distinct. Key criteria for both processes should include:
  (*i*) partnering and teamworking, and
  (*ii*) evidence of skills ability (for contractors).
(*d*) The client department must lead the project at all times, even after the appointment of a client adviser, project manager or other consultant.
(*e*) A value for money framework should be established for each project which ensures a structured approach to planning and managing a project from inception to completion.

Once the client has picked a designer, then that designer should sensibly be allowed to prepare a suitable design. This requires time. Hasty preparation almost invariably leads to cost overruns later. These may be due to variations, inadequate design details and extensions of time. Adequate time to enable thorough preparation is rarely time wasted. It will facilitate a better design, reduce the likelihood of cost overruns and make for a safer construction project. Time is a resource as defined by the CDM Regulations. It must be adequate. The Egan Report (1998) recommends adequate 'up front' resourcing.

Adequate funding is necessary and funds need to be available at the right time if the client is to obtain maximum benefit from a project. We found that adequately funded projects tended to go well while those not so adequately funded progressed not quite so well. There was never enough money to pay for anything but the cheapest solution. The definition of time as an element of resource means, that price itself must be adequate.

Clients running an existing business and considering a project which will affect that business, will usually need to minimise the effect of the one on the other. The intended project will need to be prepared in a way which does this. Thorough pre-planning and a careful consideration of health and safety issues is needed. The client should sensibly have decided his use for the project on completion and ensured that the design team is aware of this. The costs and needs of the project in use can then be considered as part of the overall design proposal (see Latham (1994) and Egan (1998)).

Who should the client pick as his principal contractor? Certainly one he can trust to do the job as required. If often helps if the contractor is known already. He needs to have adequate and relevant resources available to do the work when it is required. Common sense, partnering and CDM all help selection of the best option. Considerations for a successful client/ principal contractor relationship include:

- the appropriate management skills and experience relevant to the job — many managers openly admit to lacking knowledge of the work which is carried out by specialist sub-contractors, yet they have to manage such work, often on a major scale
- fair conditions of contract, preferably standard —start on a basis of trust
- trust and confidence between the parties
- a sensible programme — this is not necessarily fast, nor is it slow — it is the programme required to do the job properly
- the best available resources — of all types. Operatives and tradesmen who are knowledgeable of the demands of the type of work. The need to be available in the requisite numbers and skills for the requisite time
- the plant provision needs to be fully adequate, i.e. able to provide the required service to the required points at all times. Avoid any inadequacy or unavailability. These usually cost money
- the supervision provided must be adequate. The increasing use of sub-contractors needs to be matched by an increase in supervision numbers — a factor often not appreciated.

Before the project is allowed to commence, the principal contractor must prepare his construction phase health and safety plan. This takes time. As the client is not empowered to order commencement of the work until the plan is approved, it follows that starts to construction work as soon as contracts are exchanged are in breach of regulations. The contractor needs time, not just to prepare his health and safety plan, but also to secure his resources, place orders and plan the project as appropriate. These tasks are carried out far better before the start of the construction work itself. Once work commences on site, other pressures take priority. Failure to complete the necessary pre-start tasks tends to create difficulties. It is often against the client's business interest to insist on a premature start. Yet many clients do just that.

The client, in setting out his requirements for the project, and appointing those organisations he feels are best equipped to execute the project, will sensibly ensure that good communications exist throughout the team. The communications should tell *everyone* what is required of them and be presented in a way which enables everyone to easily understand what they have to do and make it easy for them to do it. If the client can achieve this then he is optimising his chances for getting it 'right first time', of getting a suitable project at a sensible price. By the same token, the project team and individuals within the team will have contributed positively to the project and, hopefully, each will have acted in a similar manner.

The client will often have his own project team to help him make the necessary selections and to monitor the project through to completion. The appointment of a capable planning supervisor, as required by the CDM Regulations, can be of major assistance *provided* that person has experience which is relevant to all aspects of the project and is not simply a CDM administrator.

Procurement experience of university and further education college clients indicates that, where there is a clear commercial payback, then private finance, construction and operation comes into its own and much student accommodation is currently being provided using the private finance initiative (PFI) approach. Most projects are, however, one-off and little need is seen for long-term framework agreements.

There is, now a move towards two-stage tendering. Clients are keen to define designs which optimise buildability and minimise lifelong costs. They get best results by involving contractors and suppliers as early as possible in the design process. This is felt to have improved quality dramatically. Everyone is regarded as a consultant and their early views enable 'best practice' to be incorporated in a proposal from the onset. The requirement to 'sign off' information makes people much more careful as they finalise their proposals.

Construction tenders are only invited when costs and designs are largely finalised. Those involved in the design stage will not necessarily be involved in construction. Awards based on lowest price are, however, becoming fewer and more emphasis is being placed on price negotiation to get best value.

There is increasing emphasis on quality rather than cost. The management of change and risk is scrutinised carefully, the parties sharing pain as well as gain. It is logical for any client to allow a contingency for client-led variations. By the same token, designers are felt responsible for design-related construction problems.

The Association of University Directors of Estates (AUDE) is looking to make risk sharing common practice and to give contractors an incentive to reduce costs and programme times by sharing the savings made.

Bidders for projects are being selected increasingly on the basis of relevant experience, safety, and company culture. The defect liability period is set at anywhere between one and five years, the contractor being made responsible for operation during this period. Latham refers to liability post-completion.

A further initiative by AUDE relates to training. Bidders are asked what training they carry out, what further training would be required if they won the job, and what training they would implement if the client paid.

The Highways Agency had four Managing Agent Contracts (MAC) under way in the summer of 2002. They cover the term maintenance of trunk roads and are expected to lead to better value and quality of service. Audits and measurement of performance form the basis of the arrangement. Suppliers are expected to benefit from increased profitability. Later benefits will arise when national indicators have been developed to benchmark and score the Highways Agency and its contractors. While earlier term maintenance contracts were for three years, the new arrangement is for five years with a possible two-year extension. Under the terms of the MAC, all maintenance work up to a value of £500 000 is carried out while the term contracts had a maximum value of £250 000. The MAC is also responsible for the design, procurement and supervision of discrete improvement schemes up to £1 million and major maintenance projects of any value. Features of the arrangement are transparency is open book accounting. A strong system of auditing is required. Bid assessment was on the basis of 80/20 weighting of quality and price. Partnering arrangements are being developed in the supply chains and a system of performance measurement is being developed. The new arrangement seems to be working very well.

The Environment Agency, working in a similar way to the Highways Agency, has split the UK into a series of regions and appointed consultants and contractors to work in each region on an exclusive basis. This is enabling teamwork to commence at the onset of a project, enables established teams to be maintained, and minimises wasted effort.

Marks & Spencer Ltd have used the area contractor method most successfully for many years. The client gets what he wants, when he wants it, without excessive disruption and to a high quality. Contractors work as part of an established team, have an assured workload and see continuity and a sensible financial return. Transparency and open book accounting are central to the arrangement.

Architect/designer

Ian Ferguson (1989), in his book *Buildability in Practice,* defines buildability as 'the ability to construct a building efficiently, economically and to agreed quality levels using the various materials, components and sub-assemblies'. To this we would add that construction work be designed to enable safe and cost-effective construction, maintenance, alteration and demolition to take place. This requires an awareness of all elements of project work, from inception to demolition and what those involved in each element require to work effectively and safely. In essence, what details should be put on the drawings to make the work of the steelfixer, the blocklayer and the maintenance engineer, for example, easier to carry out?

Easy work will tend to be cheaper and quicker to execute than difficult work, so satisfying key client concerns. It will also tend to be safer — by helping to reduce time spent working at heights, for example. This requires experience of the types of work involved. Contract debriefs held at the end of each project and listing, among other things, examples of good and bad construction practice, can be used to accumulate a list of 'good to have' construction features. This is communication from the end user to management. A factor often ignored in business. Egan (1998) talks of designs for construction and use. Latham (1994), in considering professional education (see section 7.26 of the Latham Report) asks whether professional education needs a greater content of practical experience and refers to previous proposals on the topic. Appendix V in the Latham Report gives recommendations to the professional institutions of the construction industry (the *Crossing boundaries* report by the Construction Industry Council (CIC, 1994)). Multi-disciplinary education and skill provisions feature strongly in those recommendations.

Lord Robens, chairing a committee considering health and safety issues for the government in 1972, commented: 'The primary responsibility for doing something about the present levels of occupational accidents and disease lies with those *who create the risks* and those who work with them'. Avijit Matra (1999), writing in *Civil Engineering*, noted that: 'Every time an engineer puts an idea on paper, he or she may commit people to working in a hazardous environment; that is, he or she creates a risk. In spite of legislation which requires the provision of measures to control these risks on site, the incidence of accidents is construction is still far too high'.

In many cases the accident could have been avoided with more thought in the design stage. The report *Safety and health in the construction sector* (Commission of the European Communities Directorate) indicated that in 60% of cases more could have been done at the design stage to eliminate the hazard. Designers could eliminate unnecessary hazards. This point is addressed by the CDM Regulations.

Managers with site experience will insist that safety and the degree of buildability are interlocked. You cannot have one without the other.

(*a*) The architect/designer, on behalf of his client, has a key role in producing a good design. It does not necessarily involve high cost. Good design will provide value for money in terms of both total cost and cost-in-use. The energy and maintenance equations should be uppermost in the minds of the client and the designer, as well as the appearance of the façade and the effective use of space (see section 1.18 of the Latham Report).

(*b*) The elimination or reduction of risk will increase buildability and reduce costs. The production of hazard lists, the risk assessments on them and the actions taken to eliminate or reduce those risks, can form a library of good practice for the future. We will not need to be forever reinventing the wheel!

(*c*) We can communicate buildability by the specification — by specifying sensible tolerances, or readily available and safe materials, or stating sensible programme requirements. The drawings, indicating how the structure will be put together, will ideally communicate that same buildability. If it does, work will be easier to carry out and cheaper. If it does not, then the opposite will apply.

(*d*) Partnering will enable early input from the contractors involved. That input will largely be about simplifying the construction process. It will be based on practical experience of the job — what works best.

The main contractor

The main contractor is responsible for, among other things, making a profit. In common with all businesses, he will only survive if the business is profitable. While it is not essential to make a profit on every project undertaken, it will certainly be the intention to do so. He will expect reimbursement for any justifiable additional costs. Clients need to be aware of this. Trust between the parties is likely to assist in settling any problems which occur.

(*a*) The main contractor is likely to happily work with standard forms of main contract, but could well find problems in accepting altered versions. Clients can help by using standard forms as Latham (1994) suggests.

(*b*) Spending a relatively large sum of money in carrying out the construction of a project. It is essential, from the client's point of view, that this money is spent well. The client will want to minimise any cost overruns and only to pay for genuine extras. If the two parties have worked well together previously, then this might well be more advantageous than the lowest price alternative from a contractor not well known to the client.

(*c*) Effective planning of the proposed construction work and the procurement of resources. This takes time and should be completed, as far as practicable, before construction work starts. The CDM Regulations require the completion of the construction phase safety plan, and its approval by the client, before construction can be sanctioned. The allowance of appropriate time makes business sense.

(*d*) Completing construction work in accordance with the agreed construction programme to the specified quality. This will give the client what he expects when he expects it. It will enable the client to finalise his own plans and carry them out without interference from the construction process. It is crucial to industrial, commercial and retails clients whose business continues as the contractor works around them. The avoidance of disruption is paramount.

(*e*) Provide adequate resources, suitably skilled, at the appropriate time. An ability to plan and organise correctly is needed and this requires experience of all aspects of the project. In our experience many site staff have inadequate knowledge of sub-contractor requirements. This applies particularly to specialist sub-contractors. The lack of knowledge is seen as a key weakness in main contractor organisations. Clients need to be aware of this. Proper observance of the CDM Regulations should preclude inadequate provision.

(*f*) Managing the construction work effectively. The increasing use of sub-contractors needs to be balanced by an increased level of supervision. Too many contractors fail to ensure this and work suffers as a result. Delay and disruption then lead to an increase in cost.

(*g*) Providing an effective health and safety regime on site. This includes safe systems of work, relevant training and ensuring that those who carry out work are adequately trained. The plant provided needs to be safe in use and adequate for the job it is required for. Materials must be selected, stored, handled and used in a safety conscious manner. Reputable contractors are fully conversant with the requirements and tend to work stringently to the appropriate regulations.

To work successfully, the main contractor needs the relevant information at the right time. He has a vested interest in making the work easy to construct and will contribute freely and in good measure to ensure this. This will also tend to be in the client's interest and it is likely to help designers and architects who will often have less construction site experience.

In our experience as contractors, it was found preferable to employ all staff and certain key operatives directly. The concrete placing and finishing, the site service gangs, the banksmen on the cranes, the canteen and office assistants, were all directly employed. This helped ensure that we produced a quality job. Equally importantly, we kept control. Contracts where we used wholly sub-contract services and controlled these with our own staff, were not nearly so successful.

Sub-contractors (including specialists)

Sub-contractors are businessmen and also need to make a profit. They will seek sub-contracts with clients and main contractors who provide the best opportunity for this. Important considerations for sub-contractors include the following.

(*a*) A fair price for the work, agreed without a Dutch Auction taking place.
(*b*) A fair, preferably standard, form of sub-contract.
(*c*) Fair and prompt payments for work carried out.
(*d*) A desire to simplify the work being carried out. This will cut costs, reduce the likelihood of disruption due to building difficulty, reduce programme times, and give opportunity to improve the financial outcome. When construction proceeds well, it tends to generate its own improvements as work progresses. Things get better and better. The sub-contractor will usually assist the design and main construction teams freely to get such simplicity.
(*e*) A sensible programme to work to, which allows adequate time for off-site preparation and fabrication.
(*f*) Being able to do their work without being disrupted unnecessarily by others. If successful this will help production, minimise additional cost and work to the benefit of all members of the construction team.

(*g*) Effective support from the main contractor particularly in the provision of agreed services. Offloading, storing, provision of craneage, for example. Craneage provision can be a problem. Regardless of size, each crane can only do one lift at a time, and the available hook time can be insufficient. As inadequate hook time will result in delay and additional cost, correct craneage provision needs careful consideration by those concerned from the onset.

At a time when most construction work is carried out on a sub-contract basis, it is important that clients and main contractors select their sub-contractors on a basis of performance. Failure to do this can create major cost and programme overruns.

By the same token, the increase in sub-contracting often means that the client, main contractor and design teams do not have the in-house skills which are necessary to produce the best, most buildable and economic design for a project. It is imminently sensible for the teams to get prospective sub-contractors involved at the earliest possible stage to use specialist sub-contractor knowledge to produce the best design and to maximise the benefits which flow from teamwork and communication.

Major problems can occur for the client and the main contractor if the sub-contractor fails to perform. The advantage gained by using input into design needs to be maintained until work is completed. This means, on the one hand, that constant monitoring of sub-contractor progress takes place and, on the other, that services and support are provided fully and as agreed.

A sensible main contractor will have regular progress meetings with sub-contractors. How these are arranged is not a significant issue. What is significant, however, is that they are effective — the main contractor needs to resolve problems as they occur, or, better still, prevent them occurring. The sub-contractor has identical desires. Regular and effective communication can be crucial.

Direct labour — that employed directly by the main contractor

A key point, which we made earlier, is the need to provide an element of one's own labour. To resist the desire to sub-contract everything. We include in the description of direct labour those smaller sub-contractors (steelfixers, carpenters, for example) who have ongoing relationships with the main contractor and provide regular support to the direct labour team.

Based on our own experience, we would find it difficult to recommend the employment of a main contractor who did not provide a direct labour facility. Contracts were certainly more successfully carried out with the facility than without it. At the same time, the need to train a workforce for the future (see Latham (1994) and Egan(1998)) must mean that reputable organisations in the industry will play their part in making this happen. The direct labour element tends to work very much as part of the company. It is loyal, very often multi-skilled, well known to supervisory staff, and a fundamental aid to effective teamworking on site.

Once we have selected what we regard as the best team to do the job, putting together units who can perform their various tasks to best advantage, we need to ensure that the team works effectively at all times. Teams and businesses are improved by a process of increasing their strengths and eradicating weaknesses. Strengths will logically increase with practice of doing the job, with increased practical experience. Proper team selection should give us a good start. Eradicating weaknesses requires a constant awareness of what is happening and an ability to act and rectify when things go wrong. Awareness requires good quality supervision, adequate in numbers and a reporting system which enables weaknesses to be quickly identified. We need speedy feedback.

Feedback

While management works from the top down, feedback can work in any direction. Although it is usually regarded as upward working — from site operatives along the chain of command to senior management, it can also be lateral — where relevant points are passed on by people in different disciplines or on different projects to colleagues working in other areas. It can be very usefully employed by senior management who want to keep their full team 'in the picture'.

Properly used, feedback will inform as necessary and as quickly as possible. It will put everyone in the best position to identify and eradicate defects quickly.

Senior management, seeking to build their teams, will give feedback at seminars, meetings, appraisals. Staff will be told how *they* are performing, how *their* business is performing, and the future plans for their company. This is excellent practice. It costs little and can achieve a lot. This is vertical feedback — going from top to bottom. It will be most effective if it reaches the roots of the business.

Enlightened managers will cross-fertilise feedback with managers in other departments. They learn from each other, bad practices are improved, relationships develop on a basis of cooperation and the business is strengthened as a result. Some managers, perhaps due to an inner feeling of insecurity or jealousy, fail to cross-fertilise and the business is weaker as a result. Such businesses can appear shrouded in secrecy and suspicion. This is horizontal feedback.

Vertical feedback, from bottom to top, will largely be the reporting of defects. Some businesses discourage this, feeling that it can create a forum for criticism. As a result, the defects are likely to remain.

If a business is prepared to use feedback positively, the company suggestion scheme is an example, then a lot of benefits can accrue:

- people are aware of areas where colleagues have had problems and can take steps to avoid their recurrence
- the business becomes aware of its problems at an earlier date and corrective actions are likely to be easier to implement and more effective
- inexperienced staff can benefit. Formal lists of good and bad practices — 'do's and don'ts' — will help cover gaps in individual knowledge which occur due to that inexperience. Young designers and managers can avoid some of the errors they currently make — often providing unbuildability while seeking the opposite.

Working with an eminent consultancy recently, two factors became quickly apparent:

- design engineers said that they would welcome construction progress meetings and debriefings at project end — these would provide specific information on good and bad practice and what actions to take in future
- the same engineers suggested that each drawing should be self-sufficient and not rely on other drawings for support — the tradesman could then work on site with a single drawing at any one time. The intention was to avoid errors. Management had stated that, provided the information was total, drawings did not have to be stand-alone.

Consider yourself on site, working to a drawing and probably not being aware that it was incomplete. How would you ensure that your work was correct?

Where would you look for data missing from your drawings? Would you even know that data were missing?

Architects and designers would benefit from such feedback — it would be used in generic hazard lists and as guidance for new graduates.

Perhaps the ultimate benefit of vertical feedback from bottom to top occurs due to the increasing amount of specialisation which has occurred in the construction industry. Specialist sub-contractors have knowledge which is unique to them. This needs to be passed to architects and designers at an early stage in the concept and design stages of a project. Best practice in construction can then be adopted.

An international contractor based its tender price for a project on mechanical and electrical requirements which were assumed necessary at an early stage of the design process. As the design developed the assumptions made proved inadequate and construction costs rose alarmingly. Earlier specialist input could well have helped. Lessons can perhaps be learnt for the future.

Sensible targets and sensible intentions

The recession in the construction industry, which commenced in 1988 and has only recently abated, caused serious over-capacity. There was far too little work to spread around. Businesses of all types, anxious to continue their operations, fought hard for market share. Clients, aware of the keeness of the industry for work, were able to impose their own terms and were in a position to accept very low prices. A colleague in a utility organisation told us that the tender prices for sewage works had declined by 40% in one year. As inflation remained positive and contractors had, at best, been putting in 10% margins, then this is clearly unrealistic. Perhaps worse, such price pressure probably covered most of the industry.

Those seeking work — architects, consultants and contractors — had to progressively reduce their tender prices in order to obtain a fair share of work. Main contractors, employing mainly sub-contractors and purchasing materials, drove the price reduction to the end of the supply chain. All concerned supplied goods or services at prices which were often unrealistically low and perhaps unsustainable. Failures of performance by materials suppliers and sub-contractors resulted in contra charges from procurers further up the chain. The provision of late information by designers or variations by clients attracted immediate requests for reimbursement of extra costs. Late payment of accounts and Dutch Auctions when negotiating orders became the norm.

While this is unfortunate, in the absence of a sensible tender price initially, it is perhaps necessary practice if a business is to survive. Many of the recommendations of the Latham (1994) and Egan (1998) reports are aimed at *ending* such practices.

One way of reducing tender prices was to reduce the intended construction period. This was done, not by increasing resources, but by allowing the normal resources and expressing a determination to drive them harder to complete work more quickly. A continuing downward price spiral precipitated increasingly shorter construction programmes. '*Reductior ad absurdum*' was nigh. Such reductions result in contract overruns, sub-contractors are pressed to commit increased resources when their price is probably less than adequate in the first place. Business failures occur.

Most work in the construction industry is small in nature. Many items of work — concreting, excavations, drains, bricklaying — are common to all jobs and require a particular level of resource to execute them correctly. The nature of the work lends itself to a particular sequencing of the construction process. Without a total re-think of process, such work needs to be carried out using similar resources and similar programmes, i.e. you use sensible levels of labour and plant to achieve a sensible rate of progress.

The doubling of resources is unlikely to reduce the programme time by 50%, while the halving of resource is likely to prolong work interminably. If we are to use existing technology, then a sensible level of resource is needed to carry out the work in a sensible time.

While the sensible level of resource and time will vary from job to job, it follows that each job will have its own 'best' programme. A single-span bridge over a road or stream takes about 21 weeks to construct, times for traditional building work are predictable, a concrete box reservoir will take around 26 weeks, and so on. Attempts to carry out standard types of work to programmes much different to the norm should be scrutinised carefully.

Major projects and special 'one-off' jobs can, and do, vary from the norm insofar as programme is concerned.

A motorway project will usually contain several similar bridges. While they may be constructed in sequence, there is likely to be similar work going on in several areas at any one time. The drainage, structures, earthworks and carriageway are likely areas for this to happen. While there is now more opportunity to reduce overall construction time, there is a limit to the degree of reduction.

A special 'one off' job can be treated as 'fast track' or whatever is desired. By being special it can be designed from the onset to achieve the desired result. If, however, it is considered as a normal construction job, it is unlikely that programme timing will change much, despite our best endeavours.

Resources can, and are, increased more easily on large rather than on smaller jobs. There are, however, limits to the resources available and the resources which can be deployed efficiently on any site. This level needs consideration from the onset.

The obvious conclusion, which most are already aware of, is that each job has its own sensible level of resources and programme time. With the best of intentions, it is unlikely that much sensible improvement can be made.

In the CDM Regulations, Regulation 9 requires the allocation of adequate resources to health and safety and these resources will be competent (Regulation 8). Resources are defined as including the necessary plant, machinery, technical facilities, trained personnel and time. This implies an appropriate programme and resources to do the job safely. This conforms to the intentions of the experienced engineer who will provide the same resources for the same time.

Consideration of programme includes deciding when to commence work. What is the start date? A contractor will invariably prefer to delay a start to work on site until planning and procurement is completed (on small projects) or well under way (large projects). Work can then start in a planned manner and tends to progress better than is the case with a premature start. The CDM Regulations (Regulation 4.17.d) impose a duty on clients to ensure that the construction phase of a project does not start until an adequate health and safety plan is in place. On grounds of regulation, practicality and common sense, therefore, work should not be started on site until the appropriate time. Yet many clients are still requesting a start be made on site directly after the placing of their order for construction work.

Effective resourcing

The resources used on site projects are traditionally described as:

- labour (including staff)
- plant (all plant and equipment in use on the project, whether owned or hired)
- materials (for permanent and temporary works)
- sub-contractors.

Labour

While each company has it own approach to the employment of direct labour and staff, differences in employment are largely of detail rather than substance.

Staff members will tend to be paid on a monthly basis, direct labour operatives on an hourly basis. The operatives will be paid at rates equal to, or above, the agreed rates described in the Working Rule Agreement (WRA). The WRA sets out the minimum terms of payment and working conditions. The rates and conditions are agreed and updated regularly by a working party drawn from members of the industry's trade unions and the Employer's Federation.

The law does not differentiate between hourly paid operatives and staff. Companies are responsible for initiating their own staff policy and rules. The law on pensions is the same for staff as it is for operatives, and employees can choose whether or not they wish to be members of the employer's pension scheme.

The law requires the Statement of Particulars, issued to employees under Section 1 of the Employment Protection (Consolidation) Act 1978, to specify the grievance procedure available to an employee. Company disciplinary rules must also be specified.

Training will be provided. Induction training is mandatory, and there will be training relevant to the safety issues of a project. The training necessary to acquaint the workforce with all aspects of their work and the use of the tools needed to carry out the work will also be provided. Other facilities might include craft and operative skills training at local further education colleges or via the Construction Industry Training Board. Many companies use in-house training to ensure that staff and employees are well versed in the particular knowledge requirement. Many employees will attend professional institution (e.g. the ICE, the Chartered Institute of Building, etc.) meetings, or some of the many public courses which are available.

In the best companies, labour and staff will be well trained, well motivated, well appreciated and will carry out their separate tasks with concentration and determination. This may sound rather trite — it is, nonetheless, true.

Labour is a key resource. Recession in the industry has reduced the numbers available and there is a shortage of key skills. This applies to both the main and sub-contract organisations. This shortage needs to be recognised from the onset of a project.

With less skills training, fewer craftsmen, higher wages and tighter construction programmes, designs need to be prepared in a manner which recognises the problem. We need to increase buildability and should consider actions such as:

- maximise the amount of off-site assembly and prefabrication — this will also tend to improve quality
- reduce work carried out on site as much as possible

- standardise components as much as possible
- ensure that the plant and equipment provided is fully effective
- ensure that sub-contract labour can and does perform
- weatherproof the site if possible — this will reduce downtime, increase production and improve quality
- avoid wet casing as fireproofing, use dry or spray-on casings
- maximise the use of plant and provide the most effective tools available — use low-weight tools, preferably cordless hand-held items, and equip the workforce extensively with the appropriate tools
- minimise double handling on site by careful planning and arranging 'just in time' deliveries where appropriate
- package material deliveries where possible and use forklift trucks to load directly onto scaffolds
- use knock-down kits for specialist fittings or have them pre-assembled
- prepare and approve method statements at an early stage and ensure everyone involved is aware of the intentions
- ensure that those preparing the statements have the relevant practical experience (and avoid trial and error to get it right)
- ensure that setting out is always accurate and so avoid 'fudging' work to correct errors
- a simple design will be easier to construct than a complicated one
- maximise the amount of work you can do under controlled conditions (avoid dependence on weather, for example)
- allow sufficient time to do work properly.

Plant equipment and tools

In our experience projects progressed well and made money where plant was deployed well and performed to its capabilities. On the other hand, we tended to get poorer results where the plant was restricted, either in use or provision. With restricted labour availability, effective and adequate plant provision is essential.

Plant usage needs proper planning — good crane hardstandings and haul routes are key issues. Scaffold needs to be effective in width and positioning. Width assists effective working. The gap between the scaffold and the building façade may be crucial to position façade elements. Portable tools should be of minimum weight. The effectiveness of a compressor is often governed by the quality of its airline hose connections or the point on the breaking tool.

Effective craneage means adequate hook time for the various site needs. On most occasions you should have a small over-capacity.

For effectiveness, plant must be properly maintained and safe in use. Site and company control procedures should ensure this. Effectiveness is assisted by the following.

(*a*) All plant must be safe and in good order. Operatives must be properly trained. Plant must be placed on safe standings at all times, and safe access and egress assured.
(*b*) Ensure plant is never undersized. Inadequate plant makes everyone struggle to do the job. The saving on plant hire by using undersized plant is more than offset by extra cost elsewhere. Your plant and equipment must be adequate.
(*c*) The key factor with craneage is hook time. No matter how big the crane, it can only do one lift at a time. An efficient job will always have adequate hook time. Make sure this includes your contract.
(*d*) The efficiency of plant — compressors and excavators, for example — is as good as the quality of tools that they use. Use the correct excavator bucket, ensure that it has good teeth and can *comfortably* dig to the required depth. Ensure compressor hoses do not leak air at the connections and that the jackhammers have good sharp points, not blunt rounded ones.
(*e*) In water-bearing ground, ensure that your pumping arrangements are well thought out. Ensure that you dispose of the water adequately. If you do not, you will simply recycle the water and never win.
(*f*) Ensure that no old basements underlie the site — their roofs can collapse under loading.
(*g*) Do not forget the maintenance of your plant. Ensure the fitter visits the site regularly. Keep an eye on this all the time — it often gets neglected.
(*h*) Ensure that the protective clothing you supply is fully effective. Cheap items never seem to last and are not respected.
(*i*) Small tools, especially battery operated ones, have a habit of getting lost. Have a system for booking such items onto site and back into the stores. Do not leave them lying around.

In equipment terms, the following points are worth noting.

(*a*) *Pump hoses*. Check for any damage prior to returning to the hire depot. Damage charges seem to be the norm.

(*b*) *Sheet piles*. Check for any damage, make sure you only pay for the pile weight, not the all up weight (including earth). We have known it to happen.

(*c*) *Trench sheets*. Note the length of the sheets and their condition when they arrive on site and when they *leave*. Note the hire and damage charges in respect of this. We have seen designs for trench support systems using trench sheets which were fully adequate when in position and fully supported. The sheets themselves, however, were insufficiently stiff to be driven into the ground and buckled under driving loads. Avoid this. Always use a good, heavy sheet. It will last longer and utilimately prove cheaper. It will also be safer.

(*d*) *Scaffolding*. Fittings tend to get lost on a massive scale. Try not to hire the equipment piecemeal. Rather let it as a sub-contract package.

Finally:

- always be fully briefed by the estimator, the plant manager, and consult with your site foreman before you finalise your plans
- some loads require police permission to move on the public highway — this takes time to arrange so bear this in mind when ordering and let the supplier, who is more used to it than you are, make the arrangements
- use specialists to erect and dismantle specialist plant and equipment (large cranes, scaffold, etc.)
- the plant delivered to site is sometimes not of the standard we expect, e.g. noisy, dirty exhaust smoke, poor equipment. Do not tolerate this. Send it straight back to the supplier. You can get first-class equipment for the same price
- beware of creating pollution of ground or watercourses. Diesel contamination can penetrate certain types of plastic water pipe — then you do have problems.

A clear control on plant is whether you hire it internally (from your own business) or externally (from other suppliers). Clear guidelines are required and should be available in an organisation, especially if quality assurance procedures are in place. They should state what your responsibilities are and how you should discharge those responsibilities.

You may wish to purchase. This will require finance and the decision will very likely depend on the capital expenditure provisions made within the budget. We tended to purchase very high utilisation items, e.g. cars and vans, compressors and small tools. We also purchased used cranes where we could see a substantial use and a well written down purchase cost. We would not purchase plant with a high attrition rate (excavators) or specialist items. Sheet piles, while they do not deteriorate when standing, can tie up a lot of money and not get used much. Each item needs separate consideration and this must be in the light of the existing business circumstances. Do not waste time requesting permission to purchase anything and everything. When you do have a good case, prepare it thoroughly.

A second type of control is the environmental issue — you must ensure that diesel leaks do not occur and cause contamination; that oil does not seep into a watercourse. Shotblasting can give problems of contamination, especially when working over water. You would be well advised to seal off the area. One factor often not expected is the very high lead content of some old bridge paints. Take and test a sample before you start.

Noise levels can be a problem to your neighbours and so can working hours. The tower crane must not oversail other properties without permission.

Your stated procedures will include various recording procedures. They are usually completed weekly and hire charges are checked against them. They need to be accurate and submitted in a timely manner. Suppliers will tend not to undercharge — ensure that they do not overcharge. If you have problems, get them dealt with straight away. Keep a note in your site diary. It may be of help later.

A common site failure is that of off-hired plant not being collected. Off-hire is verbal, the supplier collects it later, and site is charged to the date of pick-up. You should off-hire by telephone and confirm it in writing immediately — send a fax! This will save cost, argument, nuisance and possible damage.

Record any damaged items, you will be charged for them. Excessive charges are sometimes levied for damaged goods. Your record will preclude this. Loss or damage, or damage to adjacent properties, needs recording carefully and your insurers should be informed promptly.

A changed method of working may be adopted as the situation on site changes. For example, you may wish to use different plant or change the piles in a cofferdam. Whatever the change, make sure that you sit down and do a risk assessment and costing before you take action.

Remember that the actual output of an excavator will be different to the manufacturer's figures and probably much less due to the site circumstances. Be realistic in your assessments. You will also find that running costs and damage rates can be greater than you might expect.

Be clear in expressing your requirements. You know exactly what you want. Ensure others also know — *exactly*. It will save a lot of problems.

The true cost of broken down or ineffective plant is the cost of the relevant operation either stopping or going more slowly as a result of the problem. This may increase the overall construction period. While this might sound extreme, it is highly likely to be the case on a small, single operation, contract.

The final cost can be much greater than the direct cost of the plant itself. Be sensible. Always use first-class plant and equipment and have competent operators.

Materials

The architect or designer, selecting materials for use in the permanent works, will have made selections on the basis of the suitability, durability, buildability, availability and cost of the items involved. The issue now is to handle them effectively and this involves the construction work itself.

Materials must conform to the requirements of the specification. Proposed suppliers need to be fully aware of the requirements and orders placed for materials supply must be placed on the basis of compliance. In order to ensure profitability, the cost allowed in the tender for each material needs to be known and the orders placed will ideally be lower than that allowance. It is always advantageous to compare several prices quoted for any material before making a judgement as to which is the best offer.

Materials must be available for incorporation in the construction work at the right time and in the right quantity. Some materials (ready-mixed concrete, quarry aggregates, for example) need to be delivered at a specific rate to suit the demands of the job. If these criteria are not satisfied, the benefits of a cheap price will be swamped by the increased costs which are incurred as a result.

Storage arrangements on site need careful planning before work starts. Materials should be stored in a manner which minimises double handling and the likelihood of damage. Delivery will ideally be immediately prior to requirement to further reduce the likelihood of damage and storage costs.

Wastage of materials in the industry is probably in excess of 10%. At Mid City Place, a £46 million development, in London, Bovis Lend Lease and Stanhope plc introduced lean manufacturing techniques into the supply chain as part of their drive to improve both the delivery and quality of their finished product. All materials were delivered in accordance with a strict schedule that also controlled whether materials were delivered just in time, to the work place, or to a storage area. A support gang then delivered the materials in the correct quantity at the right time. This enabled the trade contractors to continue working at optimum efficiency, confident that the materials would be there when needed. This is best practice thoroughly planned. Wastage was found to be high. Up to 30% of waste leaving site in skips was unused material arising from over-ordering or inefficient working. The wastage factor was reducing by working through the supply chain, ensuring that only the correct quantity of materials was delivered to site at the right time, was packaged correctly and then stored correctly.

Sub-contractors

What type of project have we? Is it one where the sub-contractor has had input at the design stage and been involved as the plans were finalised? Or is it a project where no input has been provided and erection problems are more likely as a result? This could be a key factor in determining the price a sub-contractor submits. Commercial pressures may, however, force prices down to unrealistic levels. If the lowest price is accepted, a problem could develop later when losses occur.

We will assume that the suggestions of Latham (1994) and Egan (1998) have been adopted, that sub-contractor input has assisted in the design process, and that selection of the required sub-contractors is to take place.

At tender stage the various sub-contract prices for an element of work should have been checked. Are they:

- correct arithmetically? Mistakes often occur and these tend to show the overall price as less than it really is
- fully priced? Items are often left unpriced. This again reduces the price offered
- is the lowest corrected price acceptable? Does it comply with the full requirements of the terms of the sub-contract? Is the sub-contractor happy to carry out the work if the price has been modified? The main contractor needs positive assurance before proceeding to accept any offer.

The main contractor needs to ensure that all sub-contractors can provide fully adequate resources at all stages of the sub-contract works and that a competent and effective performance can be expected. Safety and quality needs to be assured. It will clearly help if the parties involved can work together as a team.

The sub-contract agreement should be agreed and signed prior to commencement of any of the sub-contract work. Failure to do this tends to pressurise the parties and antagonisms arise. It is good practice.

Before sub-contract work commences, the parties should have agreed:

- the attendances to be provided by the main contractor — off loading and craneage, for example
- the programme the sub-contractor will work to — the resources necessary to achieve this should ideally be recorded
- dates of valuation of the sub-contract works, when payment will be made and the way payment for additional work will be facilitated — it is sensible for the parties to agree that extra work is only carried out when instructed in writing
- the formal sub-contract, ideally in standard format, is signed by the parties.

Prior to commencement of work, the main contractor will sensibly:

- brief his staff on the sub-contract requirement, especially insofar as attendances are concerned
- arrange for the appropriate supervision of the sub-contractor — this can be crucial
- ensure that finished work is protected and poor work is condemned (as quickly as possible to minimise costs)
- avoid disruption and delay of either party to each other
- arrange adequate provision for health and safety, including inductions.

It is in the interest of everyone on site that each element performs adequately. The failure of one is likely to affect others. The effect will be adverse, extra costs will occur and these will be difficult to recover. The intention of the main contractor should be to ensure that everyone *does* perform adequately. The sub-contractor will ideally have:

- a sensible price which has been negotiated to enable the work to be carried out correctly.
- adequate resources, properly trained and available to carry out the work — these should have adequate self-supervisory capacity, be well motivated, and good team workers

- a full knowledge of the project as a whole and the specific requirements of the work package in the sub-contract order itself — the execution of the sub-contract work should then be pursued with total commitment and while problems may still occur, the cost of them is then likely to be minimised.

Good communication, as we have said before, is of paramount importance. A meeting pre-award of contract between the main contractor and each sub-contractor is essential if all likely issues are to be considered. It is far cheaper to sort out problems pre-start than later. It puts everyone in the picture and the site teams get to know each other. The meeting should largely follow a standard set agenda which is designed to cover all likely issues.

Site progress meetings, held as and when necessary, should also take place. Attendance should be by invitation. To be effective everyone required to be present should be in attendance and the meeting agenda needs to be fully relevant. The necessary decisions should be taken — i.e. avoid a 'talk shop'.

Successful construction

We now need to satisfy the client by achieving the set parameters — of programme, quality and cost, for example. At the same time, by satisfying these criteria, we will be in the best position to make a profit ourselves.

This is most effectively achieved by simplifying buildability as much as possible. Ideally, the client's approach, supported by practical architectural and design input, will already have simplified the workscope. Increasing the workability of the concrete mix, often by the introduction of a workability agent, can be a great help. In this context, the replacement of an element of the cement in a concrete mix by pulverised fly ash (PFA), reduces the cost of the concrete and increases workability, often giving a mix which is ideal for pumping. It helps reduce cracking by reducing the heat of hydration which arises from the chemical reaction occurring in the mix.

Ensuring that plant is fully effective in itself and is then deployed correctly on site can help achievement of the desired outcome a great deal. Effective plant deployment reduces the labour requirement. Projects where effective deployment occur tend to be more profitable than those where it is less effective, in which case additional labour has to be provided and programme times become slower.

It is common practice to reduce the construction programme in an effort to reduce costs. It will theoretically reduce the cost of labour, plant and site overheads used. If, however, we have to use more labour and plant and then work extra hours at premium rates to achieve such an outcome, then the cost increase arising may cancel out or even outweigh the theoretical savings initially sought. There is a sensible optimum programme and a balanced judgement is required to achieve this.

Adequate and effective control of the resources employed is essential to achieving the best result. The increased use of sub-contractors needs to be balanced by an increase in the amount of site supervision provided. Yet contractors often provide a reduced level of supervision and argue that sub-contractors should supervise themselves. A problem is created which site staff are fully aware of.

The reduced level of training within the industry has resulted in fewer foremen and there is a noticeable absence of those in the 30–45 age group. As a result, engineers, whose number has increased, tend to do more trades supervision than used to be the case. While technically very competent, they often lack the supervisory skills and practical knowledge necessary to perform effectively. Progress can suffer as a result.

Before construction work starts we need to ensure than an adequate period is allowed for planning of the work and the placing of orders. The site itself needs to be laid out so as to be as efficient as possible in terms of accommodation, storage, accesses, for example. Temporary work needs to have been finalised and 'just in time' (JIT) deliveries carefully considered, where required. Double handling needs to be minimised. Provision of the construction phase safety plan takes time.

Adequate resources must be available to complete the work on or before the required date. Resources must be adequate in terms of numbers and the required skills, and must be able to work the required hours for the appropriate time. The required resources need to be reflected in the tender if a profit is to be made. Site staff need to have the relevant experience and the necessary numbers to supervise the full workscope adequately. This might sound rather obvious. The fact remains, however, that projects which go wrong during the construction phase often do so because of the failure to provide adequate supervision.

The job requirement needs to be fully understood by all those involved, including the various sub-contractors and their operatives and trades people. The degree of complexity may lead to special requirements and quality standards, and tolerances may vary from the norm. People need to be aware from the onset, especially insofar as pricing is concerned. Arrange regular briefing/progress meetings.

The intended construction programme needs to be realistic and meet, or exceed, the client's needs. It will ideally have an in-built allowance for 'slippage' of time (due to defect remediation or bad weather, for example).

Everyone on site needs to be aware of the required quality standards and, more importantly, be able to work to them. The required procedures for the assurance of quality and safety need to be in place.

Supplier selection should start as early as possible and be completed before work starts. This enables site staff to concentrate on construction, gives time to assess suppliers prior to ordering, and proper briefing can take place before respective suppliers appear on site. The opposite scenario is where orders are placed at a later stage, sometimes in a last minute rush. Results tend to be less than satisfactory when this happens

Cost control measures must be firmly in place and fully effective if we are to ensure that the costs incurred are at the expected level and that we can recover any additional costs to which we are entitled.

It is at this stage, where the design becomes reality that effective communications become crucial. Errors made in the construction process are much more expensive than if the same errors are made pre-start of the work. As we have said earlier, this communication needs to be two way and extend to every person on the site. How can people know exactly what to do and when to do it? And how can the management be sure of that knowledge if they do not have the necessary communications? Meetings need to be held as necessary and when necessary, and arranged to meet the needs as they arise. Too many meetings fail to achieve their objective because they are not planned properly and have an unsuitable agenda, they do not have the appropriate attendees, or do not reach the conclusions necessary. Avoid failure.

Careful records should be kept. Not just of what we do, the weather and the resources on site for example, but also of what we fail to do, and why we fail to do it. The costs of failure are largely abortive. Perhaps they are not our fault. They need to be recovered in such cases. We have seen many instances of failure not being properly recorded and then found difficulty in obtaining fair financial recovery.

The construction team should stick to the task of construction. Whatever problems occur, they will be lessened if construction costs are minimised. This means carrying out the work in a speedy and correct manner, and using the right resources. Any testing necessary should be carried out as the work progresses and not left until the end. And at times when defective parts can be changed easily. Pipelines should be tested regularly as laying takes place and before the trench is backfilled. The snagging of completed work should be ongoing. Once work is completed, it should be protected against any possible future damage. A lot of time and effort can be wasted when tests and snagging works are left until the end of a contract.

For all partners in the project team, the overriding requirement for success is to get it right first time. By concentrating on getting it right, by refusing to be distracted by peripheral issues, the cost of any problem which occurs will be lessened, partners will work better together and for each other, and the negative effect of argument will be reduced.

Chapter 4

Good practice in the construction process

Each construction project has elements which are unique to that project. It could be particular ground conditions, contamination, access difficulties, issues of programme — indeed, any of the many issues which cumulatively form the project. Because of this it is essential that each project is planned independently. Despite many aspects of the work being standard practice, the special needs must not be overlooked if success if to be achieved.

Pre-contract work

Tender handover

The tender defines the sums of money allowed to construct the various elements of the project. It is the yardstick against which all resources used on the work are measured. An appreciation of the tender and its contents is vital and a tender handover meeting helps provide such an appreciation.

The tender handover meeting enables the estimating team to give full briefing to the construction team on issues which include:

- work scope
- contract conditions, specifications, drawings, Bill of Quantities and required programme
- allowances for site set-up (staff, accommodation, site layout, for example)
- allowances for labour, plant, materials and sub-contractors

- allowances for temporary works
- any other matters.

Too many engineers are unfamiliar with the tender process. As a result they fail to become involved in important management issues. This is detrimental to both engineers and employers.

Project start-up meeting

Once the details of the tender itself are made available, project staff can be briefed on the job requirements and what they will be required to do themselves. A project start-up meeting facilitates the process. The objectives of the meeting are:

- to brief the team fully on the necessary details
- to delegate the necessary actions to prepare for the contract
- to hand over the relevant information to let people contribute and get on with their tasks.

The meeting will normally be chaired by the contracts manager. Other attendees would include any or all of:

- the *site team* — manager, foreman and engineer
- the *quantity surveyor* — he needs to action the commercial aspects forthwith, especially sub-contractors
- the *buyer* — to be in a position to action any orders
- the *plant manager* — to take relevant action with plant
- the *estimator* — to give a tender brief
- the *safety officer* — to be informed of work details
- the *quality manager* — to evaluate quality issues and any necessary implementations.

A likely agenda will include details of:

- the contract
- the client team
- the contractor's team — on- and off-site staff are included
- a work briefing — the estimator gives a full team briefing on job aspects, particularly any unusual features or concerns
- tender build-up — the allowances made in the tender for the key resources

- Conditions of Contract
- drawing register and responsibility
- plant — date for planning schedule submission agreed
- materials — date for materials planning schedules and requisitions agreed. Buyer fully briefed on supplier and allowances made (this may occur later)
- sub-contractors — a general briefing. Each provision will be dealt with separately
- insurances
- programmes — who will draft out the required programme. What was allowed in the tender
- statutory authorities — establishing service positions, required notifications and site provisions needed
- safety — all aspects. Delegate tasks as necessary
- site security — preventing unauthorised access (especially children), immobilising plant, protecting the site itself and preventing theft.

Other preparations

When everyone has had the full briefing, the team can start preparations — as a team and not as individuals. The preparations include the following.

(*a*) Progress orders for:
  (*i*) *plant* — provide the necessary planning schedules/requisitions and liaise with the plant manager
  (*ii*) *materials* — prepare schedules and requisitions, and liaise with the buyer who will place orders
  (*iii*) *sub-contractors* — discuss preferences with the quantity surveyor who will action the orders.

(*b*) Seek information on health and safety from materials suppliers (for COSHH assessments) and sub-contractors (hazards and method statements).

(*c*) Prepare the construction phase health and safety plan:
  (*i*) obtain relevant information on health and safety
  (*ii*) carry out hazard checks and risk analysis
  (*iii*) complete a fire safety plan and security plan
  (*iv*) carry out the COSHH assessment
  (*v*) take whatever other actions are necessary to ensure the health and safety of everyone on site.

(*d*) Prepare relevant programmes:
   (*i*) the *tender programme* will have already been drawn up — this will ideally be resourced in a manner which complies with health and safety, while the labour and plant resources should match the tender allowances
   (*ii*) the *site construction programme* will be prepared and agreed with the engineer — it will be based on the available information and will very probably reflect the tender programme.

(*e*) A *target construction programme* may be prepared. This will be the programme you will work to on site. It will be of shorter duration than the site programme. Its intention is to ensure that you keep ahead of the stated programme intention. By managing everyone to successfully work to this programme there is every opportunity of enhancing profitability.

(*f*) Prepare a quality plan in organisations employing techniques of quality assurance.

(*g*) Agree circulation of all information, including that to and from the client, and delegate duties to ensure all information (including drawings) is kept up to date. Superseded information must be marked as such and withdrawn from circulation

(*h*) Develop any method statements and receive such information from sub-contractors.

(*i*) Tell your own people what their tasks will be. They can then start their planning and organising work. Delegate as much as you sensibly can. The foreman can often cover safety issues as well as labour, plant and outputs.

(*j*) Prepare the site security plan.

(*k*) Agree responsibilities for keeping the site diary, materials and plant returns, labour records and other internal procedures of the organisation.

Your first target must be to provide *all* relevant information to *all* the team. This will give you the best opportunity of planning the job properly and getting off to a good start.

Site establishment

This covers the requirements of site supervision and administration, working from offices located on site. It includes any or all of the following.

(*a*) *Site staff*. Agent, foreman, engineer, etc. on-site full time. Plus those visiting the site, e.g. quantity surveyor, safety officer and contracts manager.
(*b*) *Staff cars*. As the company allocates them.
(*c*) *Attendances*. Labour, and any necessary plant, to attend on sub-contractors and your own needs (dumper driver, canteen, cleaning and chainboy).
(*d*) *Travel time*. To get your own labour force to the site and to provide transport as necessary.
(*e*) *Plant* items not already in the priced Bill — crane visits, pumps, small tools, small concrete plant (mixers), etc.
(*f*) *Scaffold* where not already in rates. For large contracts it is often better priced here.
(*g*) *Offices, canteen, stores, toilets and general welfare*. Transport to and from site, hire, furbish as necessary. Includes all consumables; number and size of offices allowed is stated.
(*h*) *Laboratory*. Equipment is expensive. Will probably also require staffing for test purposes.
(*i*) *Site compound*. Provide fences/hoardings, gates and hardstanding. Remove on completion.
(*j*) *Signs*, setting out company contract, statutory notices, instruments, profiles, tapes, pins, etc.
(*k*) *Temporary works* of a major nature; cofferdams, access roads.
(*l*) The *storage area* needs of sub-contractors' materials must be identified. We often fail to do this.
(*m*) With *safety* in mind, designated accesses must be safe and hardstandings adequate for the intended purpose. Provide them adequately!
(*n*) You need to be aware, for practical, contractual and safety purposes, of any effect *adjacent undertakings* may have on you, or you on them. A minor dust problem on site, for example, can be a disaster for a shopkeeper or food retailer. Vibration from piling equipment can create problems in many ways.
(*o*) *Visit* the site itself with members of your team.
(*p*) Can you *improve* any of the proposed methods? When space is tight you have to make compromises. Prepare your plans to give adequate space to do the job properly as a first priority. You may then find alternative personnel, accommodation areas, etc. We have seen too many jobs where accommodation areas were sensibly set out but the site struggled — there was too little space for other needs.

Temporary works

Make sure the team fully understands the construction requirements, especially the temporary works elements. Can the installation methods be improved? Problems can be encountered if trench sheets, which are of too light a section to allow proper driving, are used.

(*a*) Ensure that the site foreman is satisfied with the equipment you intend to use. If he feels that a different item would be beneficial, listen to him. A little extra initial cost can be very beneficial if you get full adequacy as a result. Never let a job struggle by using inadequate resources.

(*b*) Have a pre-contract meeting with each proposed sub-contractor to ensure that each party is fully aware of, and satisfied with, the contract and construction requirements.

In short, if there is going to be a problem, now is the time to sort the matter out. It is far cheaper now than later.

(*c*) The period post-contract award and pre-contract start is the best opportunity managers have to ensure contract success. Anything you get wrong now is likely to be more difficult and costly to rectify later.

Your need to get it right first time. That means now.

Demolition

Guidance Notes G729/1–4 (HSE, 1988) are the response of the Health and Safety Executive (HSE) to the need to assist industry to ensure safe demolition procedures and to comply with the law. An accident during demolition work is much more likely to be fatal than on other types of construction work.

Accidents are generally due to premature collapse or falls from height. These are often a result of poor pre-planning. This leads to site operatives devising their own methods of access and methods of work without full information on the dangers inherent in the demolition process:

- G729/1 covers preparation and planning
- G729/2 covers legislation
- G729/3 covers techniques
- G729/4 covers health hazards.

At tender stage

(*a*) Tenderers have sufficient information to prepare the tender.
(*b*) Clients provide details of the structure, its construction and previous use. This enables the danger from hazardous substances to be assessed and to decide the best methods of carrying out the work. Hazardous chemicals or radioactive materials may have been stored. The structure itself may be contaminated.
(*c*) If information is inadequate then a structural survey should be carried out by a competent analyst.

The demolition survey

(*a*) Prospective contractors *must* ensure that they are provided with sufficiently detailed information to identify possible problems with the structure or hazardous/flammable substances.
(*b*) The survey should take account of the following:
    (*i*) adjoining properties (do they get support from the structure to be demolished?)
    (*ii*) the type of structure and the key elements in it
    (*iii*) the condition of the elements
    (*iv*) any requirement for temporary works or staging during demolition
    (*v*) are there any confined spaces?
    (*vi*) are there hazards from asbestos, lead, contaminated land, etc?
    (*vii*) is access and egress adequate?

Preferred method of working

The preferred method of working during the demolition process is as follows.

(*a*) Demolition of a property is usually carried out in the reverse order to the construction. Normally the building height is reduced gradually or a controlled collapse is carried out to allow work to be continued at ground level. The intention of this is to stop people working at heights.
(*b*) Shears, hydraulic arms and balling machines can assist but you must ensure that there is sufficient area for their use and the plant itself is adequate for the task.

(*c*) Where working from the structure is not possible, elevating working platforms or safety nets or harnesses may be used. Any net or harness must be properly secured.

Method statement

A detailed method statement, prepared before the job starts, is essential. The method statement should cover:

- the sequence and method of demolition noting access/egress details
- pre-weakening details of the structure
- personal safety, including the general public
- service removal/make safe
- services to be provided
- flammable problems
- hazardous substances
- the use of transport and waste removal
- identity of people with control responsibility.

Public protection

Protection of the public is of paramount importance. The following should be adhered to:

- fence all demolition work — the fence should be at least 2 m high
- do not allow demolition debris to accumulate on floors
- clear debris at ground level regularly
- fence any holes to prevent people falling
- immobilise plant when not in use — this will prevent unauthorised use
- isolate or make safe all existing services on the site
- remove access ladders and store securely — this prevents unauthorised access, particularly of children
- if demolition nets are used, keep them free of debris.

Preferred sequence of demolition

(*a*) Remove toxins, asbestos, etc., first.
(*b*) Determine demolition sequence based on the design of the structure.

(*c*) Allow regular clearing (do not allow floors or structure to become overloaded).
(*d*) Remember that, as demolition progresses, the structure will become, in most cases, less stable.
(*e*) Parts of floors may be removed to permit debris to fall. When this occurs, or when we carry out partial demolition for an alteration project, we must seek competent guidance on the resulting structural condition.
(*f*) Chutes may be used to discharge into a hopper. Only those involved in the demolition should be on site. Any asbestos sheeting being removed should be handled with care.
(*g*) Much of the corrugated metal, plastic, or asbestos sheeting on roofs is fragile and of poor load-bearing capability. Use crawling boards or similar for access purposes.
(*h*) Contaminated material, asbestos, toxic or radioactive items should be kept separate from other materials and marked for special disposal.
(*i*) When demolition of a steel-framed structure is carried out, non-structural material should be removed first. Members being cut or unbolted should be supported by crane. Beware that stress release during cutting may cause the element or the structure itself to shake.
(*j*) Demolition balls are best used on a drag-line crawler machine with a lattice jib. Ensure the equipment stands on a firm, level base. Check that the jib is adequate for the ball being used. As the demolition balls puts a lot of stress on the crane, use the minimum effective size of ball.
(*k*) Slewing a machine and ball to effect demolition puts a very high stress on the jib and is best avoided. If slewing is carried out, check the jib daily.
(*l*) Any pre-weakening of a structure needs careful planning and analysis. Indiscriminate cutting until collapse occurs cannot be considered.
(*m*) Demolition of any type of pre-stressed structure should be supervised by an engineer who is experienced in demolition and has knowledge of pre-stressing work. It is helpful if 'as built' drawings and design calculations can be made available.

Considerations for the designer include the following.

(*a*) The structural deterioration of older structures. The materials used in their construction may have been weakened by age, alternate wetting and drying, or heating and cooling.

(*b*) Demolition work carried out as part of structural alterations might require a piecemeal approach — a progressive system of removal and support provision to ensure that overall structural integrity is not compromised.
(*c*) The possible support provided to adjacent structures by the structure to be demolished.
(*d*) The possible weakening of a structure by the demolition process.
(*e*) The possible contamination of local trade by the demolition process — food and electronics businesses could be seriously affected.
(*f*) Maintain local and overall stability.
(*g*) Understand the structural load paths.
(*h*) Allowable floor loads — demolition rubble and standing plant.
(*i*) Consider the behaviour of wall supports, tension members, cantilevers, brickarches and below ground cellars.

Excavation

(*a*) When planning an excavation, the depth and type of ground must be considered in deciding whether to support the *side slopes* of the excavation or to let them stand independently. The decision will be influenced by the presence of surface buildings or other features.
(*b*) If the excavation is for a structure, *working space* needs to be allowed around the perimeter to facilitate formwork erection, steelfixing and perhaps a scaffold system. A realistic minimum working width is around 1 m. In the case where excavation is within a cofferdam, remember the intrusion into the excavation of the struts and walings which support the cofferdam. Consider their effect when settling the working space requirement.
(*c*) Excavations are '*bottomed up*' using a combination of labour and excavating machinery. Always ensure that the excavator is fully adequate to dig at the stated depth, this makes the trimming of the excavated formation much easier.
(*d*) With sloping formations, always commence excavation at the lowest point and work from that point. This helps you to deal positively with any groundwater from the start. The sump needs to be sufficiently deep to reduce the water table to below the formation level before pumping starts.

(*e*) Where water ingress is high, particularly in marine or river cofferdams, use a herringbone system of porous pipes, set in trenches and placed below the surface of the blinding and leading to a sump at the lowest point.
(*f*) Use a 100 mm diameter porous pipe for the herringbone system and allow a depth of about 300 mm to the pipe invert at the point *furthest* from the sump. This will ensure adequate drainage at this point. (On really large areas use a 150 mm diameter pipe.)
(*g*) Allow a reasonable gradient on the pipes falling to the sump. In most cofferdams a fall of 300 mm will be ample. This determines the sump depth as being around 750 mm. The crucial point is that an inadequate sump or pipe system will be ineffective.
(*h*) Fill the drainage grids containing the porous pipe with clean stone, say single size 20 mm chippings, to assist continuing drainage.
(*i*) In deeper, more confined excavations, electric pumping may be preferable. The need for round-the-clock dewatering can give rise to maintenance and noise problems, both of which are minimised by the use of electric pumps. A further factor is that the electric pump *pushes* water, the impeller being submerged. A diesel pump *sucks* water to the impeller, then *pushes* it to the point of ejection from the excavation. Electric pumps always seem to give better results. Diesel pumps are bulky and fumes can be a nuisance in restricted areas. A point of which to be aware is that the large electric pumps need a three-phase electric supply. This can be expensive to install.

Considerations for the designer include:

- ground conditions
- ground contamination
- the rate of water ingress and the water table level
- adjacent structures
- traffic and the need for separation from the works
- the safe angle of repose of side slopes
- the need for working space around a structure
- temporary works requirements
- the affect of weather on earth works — on large projects, winter working may be best avoided
- the need for sensible tolerances for the positioning of temporary works to avoid clashes of line between temporary and permanent works. Temporary sheet piles may ‘kick’ when they hit an

obstruction during driving. This may then impinge on the permanent works

- consider possible problems to construction work in deep excavations occasioned by groundwater creating an upward pressure
- consider designing the permanent works in a manner which replaces the temporary need — using concrete diaphragm walls in basements, for example.

Confined spaces

A confined space is defined as any area or space that combines difficulty of access or egress with a lack of ventilation or the possible presence of noxious or asphyxiating gases. A deep excavation, cofferdam, basement, tank, manhole, sewer trench or unventilated room could be a confined space.

(*a*) Equipment to be made available on site:
  (*i*) gas detection equipment
  (*ii*) safety harness and lanyard for operatives working in the confined space
  (*iii*) rope and fall arrest for access by others or for use when at height in a confined space
  (*iv*) lifting apparatus for person recovery — situated clear of the confined space
  (*v*) 10 minute life saver sets for emergency use.

(*b*) Persons with any of the undernoted disabilities should not be recruited for work in confined spaces and persons already engaged on such work should cease to be employed in this type of work, if any of the following disabilities appear:
  (*i*) a history of fits, blackouts or fainting attacks
  (*ii*) a history of heart disease or heart disorder
  (*iii*) high blood pressure
  (*iv*) asthma, bronchitis or a shortness of breath on exertion
  (*v*) deafness
  (*vi*) Menieres disease or disease involving giddiness or loss of balance
  (*vii*) claustrophobia or other nervous or mental disorder
  (*viii*) back pain or joint trouble that would limit mobility in confined spaces
  (*ix*) deformity or disease of the lower limbs limiting movement

(*x*) chronic skin disease
(*xi*) serious defect in eyesight
(*xii*) lack of sense of smell.

Employees should be medically re-examined at reasonable intervals taking into account the person's age and duties. If there is any doubt about the fitness of an individual for confined space work, specialist medical advice should be sought.

An employer must provide every enclosed workplace with sufficient ventilation of fresh or purified air. Work study shows that the volume of a working area is never sufficient to make ventilation unnecessary.

Considerations for the designer might include:

- ensuring safety of the users of tunnels in the event of fire, the need for rapid and safe evacuation
- the possibility of industrial gases lying in low areas (cofferdams, manholes and the like)
- sewer manholes have access covers and internal dimensions to suit the sewer pipe diameter — do the access cover and manhole sizes consider safe and efficient working?
- closed top concrete box reservoirs have very small access covers — construction work is hampered as a result. Better access would reduce costs and increase safety levels
- confined spaces in structures (plant rooms, dark storage rooms) may be inefficient and space wasting — avoid the dark areas and make plant rooms bright, adequately sized and airy
- design to avoid confined spaces as far as possible.

Concreting

(*a*) Concrete consists of a mixture of stone (coarse aggregate), sand (fine aggregate), water to act as a lubricant, and cement. The cement reacts with the water and a chemical reaction (hydration) occurs. This can cause severe burns. Ensure that exposed skin is protected. Provide gloves, safety helmets and rubber boots. Many types of footwear are unsuitable for concrete work. In the event of concrete dropping into the boot, ensure that it is removed. Crippling injuries have resulted in cases where this was ignored.

(*b*) Concrete placing is very hard work. Concrete workers often suffer severe muscular and other disabilities in later life due to the unremitting effort required. The provision of a rich and workable concrete mix will make the work safer and easier. Vibration will take less effort. A better concrete finish will be provided.

Preparation for concrete

(*a*) The section to be concreted must be thoroughly cleaned. Timber or steel formwork should be lightly oiled to aid its later removal from the concrete face. The reinforcement must be clean.
(*b*) The equipment used to place the concrete needs to be in good order. The crane correctly positioned, and the concrete skip cleaned and easy to handle. Vibrators need to be tested. Have a standby vibrator in case of breakdown.
(*c*) Access needs to be tidy.
(*d*) Engineering checks are carried out on the line and level of any formwork or falsework. Reinforcement is checked for stability, accuracy and correctness of placing and fixing, and it must have the correct concrete cover to adjacent faces of the work. Re-check after concrete placing is completed. Keep watch during concreting for grout losses at joints. A carpenter can make any necessary adjustments.
(*e*) Ensure the fixity of waterbars and other such items. These are flexible and need checking during concreting to ensure that they are not displaced.
(*f*) Remove concrete splashes from adjacent work immediately. It will be very expensive if you do it later.

The concrete mix

(*a*) Minimise the number of concrete mix designs used.
(*b*) Maximise workability to make:
    (*i*) compaction and the removal of air voids easier. Air voids in concrete reduce the strength dramatically — see Table 4.1 for the air void and strength loss relationship in a concrete mix

Table 4.1. Air void and strength loss relationship in a concrete mix

Air voids: %	Strength loss: %
0	0
5	30
10	60
15	75

(*ii*) concrete flows more easily — it can be worked more easily, around reinforcement for example
(*iii*) the 'finish' on exposed concrete surfaces can be first class with a good mix.

Placing of concrete

(*a*) Blinding should be placed using timber screeds. Setting the blinding layer some 12 mm low precluded problems if the reinforcement was bent to the outer tolerance. *Never* use steel pins, driven to level, to work the blinding. This gives an inaccurate level of finish.
(*b*) With structural slabs the need is to place the concrete sufficiently quickly to prevent 'dry joints' from occurring in the slab. A dry joint occurs when placed concrete goes hard before the concrete adjacent is placed. It occurs particularly in large, thick slabs. Use a workable mix and a fast delivery (often a pump) in such cases.
(*c*) When pouring slabs, *do not* move the concrete around with the vibrators, it encourages segregation. Place new concrete in suitable volumes adjacent to already placed material and vibrate the two together, eliminating the possibility of a dry joint and making the work easier. Finish the surfaces of slabs as the work progresses and before the surface hardens. Insert the vibrator into the concrete and air is seen to be expelled. Cease vibrating as soon as air bubbles stop rising and withdraw the poker slowly to avoid creating a void.

(*d*) High-tolerance floors need careful setting up. A Bunyon Striker or similar roller is well worth consideration.
(*e*) Wall or column concrete is placed by skip. Ensure the correct skip type is used. A slow rate of pour is required and crane placing is normal. External vibrators can be useful. Place the concrete in layers from one end of the pour to the other. Layers up to 500 mm thick make positive vibration easy. In deeper layers problems can start to occur. Pouring too quickly leads to voids and honeycomb in the concrete. It can also overload the supporting formwork.
(*f*) When pouring walls or columns, have carpenters on standby to maintain accuracy. They will check line and level, and ensure no displacement occurs due to bolts loosening, etc.
(*g*) Concrete segregates when falling from heights. Use a 'tremmie tube' to prevent this.
(*h*) When using vibrators, ensure that they do not damage the face of the formwork by keeping them *inside* the reinforcement cage. If vibration is required adjacent to the face, use clamp-on external vibrators.
(*i*) Concreting in deep walls can settle after vibration. When this occurs, plastic cracking is seen at the higher parts of the wall. This can be avoided by re-vibration just before the concrete sets.
(*j*) Place concrete at night in hot countries.

Concrete curing

(*a*) Curing is necessary to help the concrete reach its design strength and to make it more durable with less surface dusting. The curing agent needs to be applied as quickly as possible after the concrete hardens.
(*b*) Curing is intended to stop the concrete drying out and to maximise the hydration of the cement. Traditional curing involves water and hessian, flooding a slab area, or using a sprayed-on membrane. The process should be maintained for several days.
(*c*) Water is the cheapest and probably the best curing agent. However, its use can lead to problems elsewhere (by virtue of it penetrating the structure) and disposal can be a problem. If hessian is used it must be kept damp.
(*d*) On horizontal surfaces, plastic sheeting, spray-applied compounds forming membranes, or damp hessian are fine.

Where further floor finishes are to be added, sprayed membranes are not recommended. In water-retaining structures (potable water), any sprayed membrane needs checking for compatibility with the specification.

(*e*) Vertical surface curing can be effected by leaving formwork in place, but this delays the contract. Damp hessian also gets in the way of further work. A sprayed compound, applied immediately after formwork that is stripped and 'rubbed down' is fine. Water curing methods are best avoided in cold weather. Immediate curing and protection with thermally insulated quilting is the best option at such times.

Concrete additives

(*a*) *Pulverised fuel ash* (PFA) is a partial cement replacement for the cement. There can be up to 30% cement replacement. The heat of hydration is reduced and there is less risk of concrete cracks appearing. The concrete itself is more workable. It is good for a pumped concrete mix. A concrete mix with PFA content is cheaper.

(*b*) *Workability agents* — a very useful additive to cement grouts which are to be placed in restricted conditions. They assist grout flow and reduce the likelihood of voids forming in the grout. We have used them for grouting up machine bases very successfully.

(*c*) *Cold weather agents* — low temperatures slow the concrete hardening and strengthening process and can render it useless. To counter this you can:

- (*i*) heat the water
- (*ii*) heat the aggregates
- (*iii*) add a chemical to kick-start the hydration.

These measures require considerable care in implementation. Take expert advise.

Completion of concreting

(*a*) Ensure cleaning up is *immediate* after a pour is completed.

(*b*) Remove any grout losses which deface walls beneath joints in the formwork, any splashes on adjacent work, or concrete spillages.

(*c*) Rub down exposed concrete walls, columns and the like as soon as formwork is removed. Pay particular attention to any lips in the concrete at joints.

Considerations for the designer include:

- improving workability as much as is sensibly practicable — use self-compacting concrete where possible
- reduce the heat of hydration to assist curing and reduce cracking problems
- PFA is a cheap cement replacement, reduces concrete costs, the mix gives lower heat of hydration levels, and is more workable
- minimise the number of mixes specified
- consider precasting:
  - use precast planks or beams in floor slabs, then add an in-situ topping
  - where very high quality finishes are required
  - where in-situ work would be slow, tedious and dangerous
  - of thin or heavily reinforced concrete sections
- avoid in-situ concrete staircases — they are expensive, slow to produce and are often of a lower quality than is desirable
- it is easier to place concrete than it is to fix reinforcement — design non-aesthetic slabs and bases to use less reinforcement.

Formwork

Release agents

All formwork should be cleaned and the surface to be in contact with the concrete coated with an *approved* mould oil prior to fixing. Follow the manufacturer's instructions when using the oil. *Do not use other unapproved oil.* Unapproved oils can have an adverse affect on the concrete.

Construction joints

(*a*) These occur at the ends of a pour and can be vertical or horizontal. Vertical joints occur between adjacent sections of a wall or slab. Horizontal joints occur between adjacent lifts of a wall or column. As a result, the joint has to provide continuity between adjacent concrete sections. Particular attention must be paid to correctly form construction joints in water-retaining structures.

(*b*) Horizontal joints are exposed upon completion of the concreting and require no formwork. They are unformed joints. These joints should be roughened by wire brushing to remove concrete laitance as soon as the concrete has properly set. It is then mechanically roughened (scabbled) prior to fixing forms for the next lift. There are few problems with these joints provided they are clean and roughened.

Some engineers prefer to put a thick layer of mortar on horizontal joints before the next level of concrete is placed. Experience shows that such mortar layers can be attacked by peaty water. Mortar layers can also spoil the appearance of finished work, as they are often a different colour to the concrete. Providing you can seal the formwork to prevent grout loss, it should not be necessary to use mortar.

(*c*) Vertical joints require forming. Expanded metal or fine mesh is advantageous where the reinforcement is congested. Left in the concrete it helps give a good bond and has no deleterious effect. Timber joints, often placed immediately prior to concreting, can be difficult to remove later. Coating them with a retarding agent helps removal but there are fears that this may affect the concrete itself at a critical point. As it is difficult to ensure full bonding across the joint, particular care is needed in joint provision. Research indications are that:

(*i*) abrasive roughening (wire brushing) is better than scabbling which can produce hair cracks in the surface and also loosen larger aggregate flakes

(*ii*) wetting the old concrete face prior to applying fresh concrete tends to reduce strength

(*iii*) wire brushing the joint while the concrete is 'green' (un-cured) gives the best strength

(*iv*) cement mortar application reduces strength

(*v*) the best results are obtained by aggregate exposure while concrete is green and then casting against a dry joint (Construction Industry Research and Information Association (CIRIA) Report 16, *The Treatment of Concrete Construction Joints* (CIRIA, 1969)).

Fixing formwork

Timber forms, glued together and screwed, give better service than nailed forms. You need to take particular care where heavy duty use occurs (high walls and external vibrators). Use lock nuts on the bolts and ensure other fixings are adequate.

Cleaning formwork

Clean all concrete spillage as it occurs. After striking the formwork, clean all surfaces thoroughly.

Check that ply faces are not delaminating.

Grout checks

Grout checks are square timber laths, equal in size to the required concrete cover. Fixed to the tops of wall formwork they are a sound feature which should always be used (e.g. 50 mm required cover, use 50 mm × 50 mm checks). A grout check ensures:

- that wall pour joints form a neat straight line along the top of a pour
- the reinforcement is given the correct cover to the face of the wall
- formwork pressure when erecting the next lift of wall is uniform.

The wall or column kicker

Kickers should be provided at all positions where concrete walls or columns spring from concrete slabs. The kicker is poured monolithically with the floor slab. The best height is 150 mm and this gives a good structural start. Lower kickers are less sound structurally; higher ones tend to slump with the weight of concrete in them. Reinforcement is usually detailed to allow for 150 mm. A correctly formed kicker:

- allows good reinforcement alignment
- assures a good joint between wall/column and base
- provides a key for the wall formwork — this improves quality and speeds construction.

Checking formwork

All forms should be fixed to the underlying concrete by bolting. The underlying concrete (kicker or wall or column) forms the fixed line at the base to which we must work. You therefore bolt tightly to this concrete line to minimise any grout/concrete losses at the joint. Check that this has been done and have carpenters ensuring ongoing fixity during concreting.

Next, ensure the two lines of formwork which will shape the wall are plumb vertical at each end.

Finally, run a string line along the top of the wall between the plumbed ends and adjust the top of the formwork to the correct line at each joint in the formwork.

On high walls, consisting of a large number of lifts, it is wise to check the levels of the top of the formwork. This prevents formwork tilting. Ensure no gaps are visible at joints. Tape floor support forms as necessary.

Timber formwork

Timber formwork is usually based on the use of the 2440 mm × 1220 mm × 20 mm thick plywood sheet.

Care needs to be taken in the selection of the plywood. Some plys, notably Malaysian hardwood, and to some extent Finnish plys, can have sugars drawn to the surface of the ply when exposed to sunlight. The sugars act as a retarding agent on the concrete face. When the formwork is stripped, the concrete face can be severely retarded and damaged. Sealing the face of the plywood with polyurethane varnish, cement grout, or a lime wash will prevent the problem. We have found varnish the best. It also extends the life of the formwork. Screw holes should be sealed flush with the plywood face.

The plywood is usually backed by 100 mm × 50 m PAR (prepared all round) timber framing. To ensure all panels in a batch are the same, the backing timber should be 'thicknessed' on the 100 mm side. This will give uniformity of depth at the joint between adjacent panels and help prevent 'lips' in the finished concrete. Backing timbers can be glued, screwed, or nailed to the plywood. The best results, and a longer life, are obtained by gluing and screwing together. 200 mm × 200 mm plywood gussets, fixed to the corners of panels, help produce a rigid frame.

Soldiers, often made of 2 no. × 150 mm × 50 mm PAR timbers some 2·6 m long, support the plywood panels against the existing concrete wall. The ply surface in contact with the freshly poured concrete should be cleaned and protected with mould oil prior to each use. Large ply-faced panels are made using a designed backing timber and proprietary soldiers, generally of steel construction.

Ply and timber forms are very flexible in-situ. They can be rapidly adapted in size to suit differing dimensional requirements. The flexibility of the materials themselves enables a better overall dimensional finish to be achieved. The high absorption factors of the ply face gives a surface finish with little or no surface crazing.

Typical problems

Table 4.2 is reproduced from *Formwork, A guide to good practice*, issued by the Concrete Society and the Institution of Structural Engineers (1986). It lists problems which occur due to bad practice and the reasons for the problems.

Steel formwork

(*a*) Steel formwork of a proprietary design and set up to the manufacturer's instructions is quick to erect and dismantle. Properly maintained, it has a long life. It is, however, difficulty to repair when damaged.

(*b*) Preparation is exactly the same as when using timber forms — you clean and then coat the surface with mould oil. It is good for multi-use straight concrete faces.

(*c*) Purpose-made steel forms are ideal for circular ground level tanks. Use one set of forms for the base slab and one for the wall above. Depending on diameter, the full central cone and then the walls above are cast complete for a small tank, or in up to eight sections to complete one large tank. It is critically important that the forms for the base slab and kicker and the walls above form a perfect match.

(*d*) Crazing or dark staining of the concrete face can occur when steel forms are used. Slight lips between adjacent steel elements at the

Table 4.2. Problems due to bad practice and their reasons (reproduced from Formwork, A guide to good practice) (continued overleaf)

Fault	Possible design deficiency	Possible construction deficiency
Dimensional inaccuracy	Excessive deflection	Metal locking devices not tight enough in column or beam clamps. Forms filled too rapidly
Joint opening and deflection of forms	Supports too far apart or section of support members too small	Vibration from adjacent loads. Insufficient allowance for live loads and shock loads
	Excessive elongation of ties, incorrect or insufficient ties	Void formers and top form floating due to insufficient fixing
	Bearing area of plate washers or prop heads/base plates too small	
	Insufficient column or beam clamps. Failure to provide adequately for lateral pressures on formwork	Plywood not spanning in the direction of its greater strength
	Insufficient allowance for incidental loadings due to placing sequences	Use of lower strength class members than designed
	On cantilever soffits: rotational movement and elastic deformation of system	Change of concrete pressure group by use of retarders, etc., or reduction in placing temperature
Lifting of single faced forms	Forms not adequately tied down to foundations to resist uplift force generated by raking props	Ties not tight enough. Ties omitted. Forms filled too rapidly. Wedging and strutting not adequately fixed
General	Props inadequate. Failure to provide adequately for lateral pressures on formwork. Lack of proper field inspection by qualified persons to see that form design has been properly interpreted by form builders	Failure to regulate the rate or sequence of placing concrete to avoid unbalanced loadings on the formwork Failure to inspect formwork during and after placing concrete to detect abnormal deflections or other signs of

Table 4.2. continued

Fault	Possible design deficiency	Possible construction deficiency
		imminent failure which could be corrected
	Lack of allowance in design for such special loadings as wind, dumper trucks, placing equipment	Insufficient nailing, screwing, bolting Inadequately tightened form ties or wedges Premature removal of supports, especially under cantilevered sections
	Inadequate provision of support to prevent rotation of beam forms where slabs frame into them on one side only	Failure to comply with recommendations of manufacturers to stand components and to keep within the limits required by the designer Use of defective materials. Failure to protect paper and cardboard forms (particularly tubes) from weather or water (or damage) before concrete is placed into or around them. Studs, walings, etc., not properly spliced
Loss of material	Ties or props incorrectly spaced, not close enough to existing concrete. Insufficient ties or props	Pies, props or wedges not tight enough Dirty forms with concrete from previous pour left on (ill-fitting joint)
At kicker	Incorrect tie, possibly causing elongation of tie. Single faced forms lifting due to inadequate anchorage. Failure to provide adequately for lateral pressures on formwork	Out of alignment kicker with stiff form Grout check omitted or placed incorrectly

Table 4.2. continued

Fault	Possible design deficiency	Possible construction deficiency
At tie	Incorrect tie	Hole in panel too large. Cones, if used, not square to panel face or not tight enough
	Insufficient panel connectors	Failure to inspect and improve tightness during pouring
Surface blemishes		
Scabbing	Incorrect release agent	Dirty forms, lack of release agent
Staining	Incorrect release agent	Incorrect release agent, over or under application, incorrect mixing of release agent
Colour difference	Wrong sheeting used, wrong treatment specification. Wrong specification of sealer for grain of timber or plywood based forms with paint, wax or similar treatment (Category 6 — see Section 3.10)	As above Form surfaces of difference absorbencies Sealanet applied to damp timber or plywood surfaces Lack of curing of concrete. Forms struck at difference times
Crazing	Very smooth, impermeable formwork surfaces may produce this effect	
Dark staining	Can be caused by use of impermeable formwork surfaces	
Between panels	Insufficient panel connectors	Badly fitting joints or panel bolts, loose wedges, metal locking devices not tightened Incorrectly erected crane handled panel of formwork

concrete face cannot be rectified. The resulting lip in the finished concrete is often very visible and difficult to remove later.

Considerations for the designer include:

(*a*) Standardise the cross sections of beams and columns as much as practicable. This maximises repetitiveness, minimises reworking of formwork and speeds construction.
(*b*) Standardise floor thickness as much as practicable. This allows standardisation of the temporary support system. The system can often be removed and re-located without dismantling and re-assembly.
(*c*) Use permanent formwork where practicable. Floor soffits and bridge decks, for example.
(*d*) Specify sensible tolerances for line and level. Fine tolerances, especially for in-situ work, can slow the work a great deal, are expensive to produce and the finish is not distinguishable by eye from that produced using a sensible tolerance.
(*e*) If fillets are to be used, at the corners of beams and slabs for example, make them a sensible standard section. Ensure the contractor uses hardwood or steel forms for the fillet. Softwood tends to produce a poor quality finish as do small sections.
(*f*) Kickers for walls and columns are integral with floor slabs and settle for a standard size (150 mm is suggested).
(*g*) Standardise floor heights as far as possible. This assists the erection of supporting falsework.
(*h*) Modularise the setting out of structures to minimise the re-working of formwork and to simplify work in reinforcement and following trades.
(*i*) Consider allowance for future needs. This usually means upsizing.
(*j*) When detailing radial reinforcement, put the laps on the perimeter, or approaching it — not on the centre. Central lapping gives rise to major steelfixing and concrete placing difficulty.

Reinforcement

The fixing of reinforcement is a task which requires much physical effort in terms of lifting, working at heights, and bending actions while fixing reinforcement in beams, slabs, walls and columns. Each of the three areas is a major contributor to accident and injury. The design, subject to the limits of the design criteria, should recognise this. Designers, carrying out risk assessments, must lessen the risk of the steelfixer as far as possible. The safer and easier the fixer's task, the less the risk. There will be further major benefits in addition. Quality will improve, the speed of fixing will improve, and construction will proceed more smoothly.

By the opposite token, a failure to appreciate the steelfixer's needs can seriously impede construction progress.

Considerations for the designer include the following.

(*a*) Starter bars. Slab reinforcement should include the provision of connecting (starter) bars at wall, column and other locations they will project from the 150 mm kicker by the specified lap diameter number (40 D, for example).

(*b*) The bar diameter — the fewer bars to be fixed then usually the safer, easier and quicker the fixing. Use larger diameters if possible.

(*c*) The weight of bars — ensure that the steelfixer can safely handle the weight.

(*d*) Minimise the number of bar marks (types).

(*e*) Vertical construction is easier than sloping work. Are tapered walls necessary?

(*f*) Eliminate any possibility of fixing errors due to a failure to clearly reflect the requirements for the top and bottom reinforcement mats for slabs, and the front and rear faces of walls.

(*g*) Ensure that the concrete mix is designed to flow easily around congested areas of reinforcement. Design such areas so as to minimise congestion.

(*h*) Consider the practicalities of a repetitive design which allows bespoke fabrication of column, walls and slab reinforcement off site, or at ground level on site, and then hoisting elements into position in completed form. This will minimise both construction time and the time spent working at heights. If bespoke fabrication is used, ensure the mats of reinforcement are rigid in themselves and suitable for lifting.

(*i*) When scheduling reinforcement, take into account factors such as the exact diameter of deformed ribbed and other reinforcement bars and the bending tolerance itself.

(*j*) Provide schedule details which will ensure adequate cover to reinforcement in the 'worst case' situation. Providing the integrity of the design is not compromised, a slightly increased cover to reinforcement will be better than a slightly deceased cover.

(*k*) It is difficult to fit reinforced concrete beams in trenches which have ground supports installed. Prefabricated beams cannot be 'threaded in' and in-situ work is made difficult if normal shear link reinforcement is used. It can help if the normal links are replaced by U-bars. An example is shown in Figure 4.1.

(*l*) Ensure that supports (spacers and chairs) are provided which are adequate to keep the reinforcement in position during concreting operations.

(*m*) When designing mesh reinforced slabs, ensure that the mesh is retained in position during concreting operations.

In terms of site management:

- keep reinforcement stocks clean and protected
- do not overload scaffolds with reinforcement
- use an adequate number of correct spacer blocks to ensure the correct concrete cover — fix the blocks securely to the reinforcement
- ensure the bar marks remain segregated.

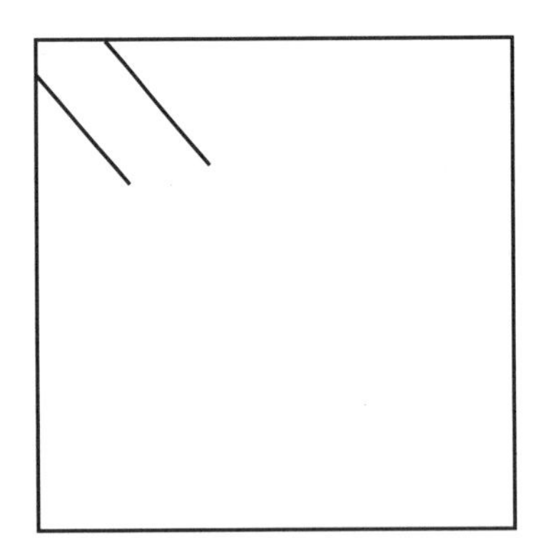

(a)

(b)

Figure 4.1. (a) Normal link; and (b) double U-bar

Scaffolding

The Construction (Health, Safety and Welfare) Regulations 1996 complete the implementation of the Directive which was started by the CDM Regulations 1994. They consolidate, modernise and simplify the older requirements and introduce some important new provisions arising from a European Directive on construction.

The main duty holders are employers, the self-employed and those who control the way in which construction work is carried out. Employees have duties to carry out their own work in a safe way. Anyone doing construction work has a duty to cooperate with others on matters of health and safety, and report any defects to those in control.

Regulation 5 — safe places of work

A general duty is imposed to ensure a safe place of work and safe means of access to and from that place of work. This regulation sets out a general requirement which applies to all construction work. It applies equally to places of work in the ground, at ground level and at height. In essence it requires that 'reasonably practicable' steps should be taken to provide for safety and to ensure risks to health are minimised. This means that action to be taken should be proportionate to the risk involved.

Regulations 6 and 7 — precautions against falls

- Prevent falls from height by physical precautions or, where this is not possible, provide equipment that will arrest falls.
- Ensure there are physical precautions to prevent falls through fragile materials.
- Erect scaffolding, access equipment, harnesses and nets under the supervision of a competent person.
- Ensure there are criteria for using ladders.

The aim of the regulations is to prevent falls from any height, but there are specific steps to be taken for work over 2 m high.

(*a*) Above this height, where work cannot be done safely from the ground, the first objective is to provide physical safeguards to prevent falls. Where possible, means of access and working places should be of sound construction and capable of safely supporting both people and the materials needed for the work. Guard rails and toeboards, or an equivalent standard of protection, should be provided at any edge from which people could fall.

(*b*) Sometimes it is either not possible to provide the above safeguards or the work is of such short duration or difficulty that it would not be reasonably practicable to do so. In these circumstances, consider using properly installed personnel equipment such as rope access or boatswain's chairs.

(*c*) If, for the same reasons these methods of work cannot be used, it will be necessary to consider equipment which will arrest falls, i.e. safety harnesses or nets with associated equipment. Scaffolds, personnel harnesses and net equipment have to be erected or installed under the supervision of a competent person.

Regulation 8 — falling objects

- Where necessary to protect people at work and others, take steps to prevent materials or objects from falling.
- Where it is not reasonably practicable to prevent falling materials, take precautions to prevent people being struck, e.g. covered walkways.
- Do not throw any materials or objects down from a height if they could strike someone.
- Store materials and equipment safely.

The first objective is to prevent materials or objects from falling in circumstances where they could strike someone. Only where it is not reasonably practicable to do so, should other means, e.g. covered walkways, be used.

All scaffolding should comply with the Construction (Health, Safety and Welfare) Regulations 1996 and, where appropriate, the Construction (Lifting Operations) 1961.

No scaffold may be erected, modified or dismantled, except under the supervision of an experienced and competent person, and all structures should be inspected by the competent person before use.

Effective scaffold erection

(*a*) The ground on which a scaffold is carried should be tidy and preferably flat.

(*b*) All standard (vertical) tubes must have a base plate.
(*c*) Where ground is soft, standards and base plates are supported on timber sole plates (often rejected scaffold boards).
(*d*) Standard boards are *mainly* 225 mm × 38 mm × 3·9 m long. Their maximum safe, unsupported span is 1·5 m.
(*e*) All scaffold requires bracing in both directions to be effective:
  (*i*) longitudinal bracing should be at 45° and at 100 ft (30·4 m) intervals
  (*ii*) ledger bracing (across the width of the scaffold) varies between 9 ft and 15 ft centres dependent on the scaffold itself. The scaffold width itself is dependent on the purpose for which it is to be used. Table 4.3 lists typical scaffolding purposes and an indication of required widths.
(*f*) Scaffolding is erected in a series of 'lifts'. A lift is the vertical distance between lines of ledgers, which in turn is where the boards to form the working platforms are placed. A standard lift is 2 m, but it may be adjusted to suit the particular circumstances for which the scaffold is required.
(*g*) An important factor is the gap between the building and the inner face of the scaffold. A gap of around 0·5 m (two boards) enables cladding panels, glazing units and other facia elements to be lowered by the tower crane into position and fixed. The working platform is then extended towards the building and boarded out to ensure safety.
(*h*) Stairways must be provided with guardrails to prevent falls.

Table 4.3. Scaffolding purposes and required widths

Purpose	Minimum width: mm	Number of 225 mm boards required
Footing only	600	3
Passage of materials	600	3
Deposit of materials	800	4
Support a platform	1050	5
Shape stone	1300	6
Higher platform to shape stone	1500	7

(*i*) Ladders must not be used if damaged, nor must they be painted. Home-made ladders must *not* be used.

(*j*) Scaffolding has to be tied to the permanent structure. Ties are fixed:

  (*i*) at a maximum spacing of 8·5 m in any direction
  (*ii*) if ties are *to be moved* after initial positioning, they must be set at a minimum of one tie per 32 m^2 of scaffold
  (*iii*) if ties are *not to be moved,* the frequency can be one tie per 40 m^2 of scaffold
  (*iv*) for the same criteria of *ties moved* or *ties not moved,* in a sheeted scaffold the maximum spacings are one tie for 25 m^2 and 32 m^2 of scaffold area, respectively
  (*v*) sheeted scaffold at height (over 50 m) requires a competent design.

(*k*) The width required to accommodate a scaffold can be 2·0 to 2·5 m from the face of the building. This is in excess of the width of many pavements in built-up areas. Permission may be required to restrict the public highway to accommodate such widths. While this may seem obvious, it is our experience that some jobs have been badly restricted because the adequacy of scaffold was not given sufficiently early consideration.

Dismantling scaffolding

While considerable thought is given to the erection process, there is often failure to give suitable consideration to dismantling — other priorities demand attention. To work safely, you must consider structural stability at all times in the following manner.

(*a*) Scaffold should be dismantled top down and not end-to-end in vertical sections.
(*b*) Ties should *not* be removed ahead of dismantling.
(*c*) Remove all parts progressively.
(*d*) Ensure scaffolding is stable when not being worked on (overnight or during holidays).
(*e*) Do not allow overload to occur by storing dismantled scaffold items on the remaining staging.
(*f*) Do not allow components to drop freely to the ground.

Considerations for the designer include:

- scaffolds form an essential means of access to structures where construction is taking place — the provision of adequate scaffold needs consideration at the design stage
- consideration needs to be given to protection of the scaffold at ground level against impact by moving vehicles — a barrier will also be required to prevent unauthorised access to site
- where scaffolding is erected on a public pavement, it is usually possible to provide a covered way at pavement level to allow continued access
- where continued access is desirable, but a covered way is not a realistic consideration, a footpath diversion may be required
- in more extreme situations, the width of the scaffold wayleave required at ground level could be:
    - structure — scaffold 0·5 m
    - scaffold width (5 board) 1·25 m
    - pedestrian barrier 0·5 m
    - diverted footpath 1·0 m
    - traffic barrier 0·5 m
    - *total width allowed 3·75 m*

  while such a width requirement may be extreme, it gives an indication of how important it can be to take such decisions as early as possible and then obtain any necessary permissions. This can take time.

Brickwork

Design and specification considerations include the following.

(*a*) Finished work should be sheeted over with polythene. This will reduce any efflorescence as well as give protection from weather and mortar droppings from ongoing work.

(*b*) Acid washing of brickwork is best avoided. It can damage the mortar and spraying may affect adjacent work. At worst, it can leak through the mortar and attack the cavity ties.

(*c*) Ensure that the same jointing tool is used throughout. Use a roller-pointing tool for raked joints to prevent damage to, and possible leaks through, the mortar.

(*d*) To ensure that there is a distribution of colour through facing brickwork, the bricks should be taken from a number of pallets at any one time and allowed to blend together. Three pallets selection is usual.

(*e*) Care needs to be taken with sample panels. The intention is to produce a panel which will be representative of the work to be carried out on site in all respects. This will not be achieved by selecting bricks in a special way for colour, etc. Such a panel will be expensive to produce and result in high factors of brick selection and wastage.

(*f*) Cavity ties in hollow walls should be fixed so as to fall towards the outer skin of brickwork. The drip should be located at the centre of the cavity. The cavities themselves must be left clear and free of debris so as to keep the inner skin of brickwork dry. The inner skin is built ahead of the outer, all excess mortar is removed, and a timber strip prevents mortar falling down the areas of completed cavity.

(*g*) The mortar used relies on consistent batching or gauging to achieve a uniform strength or colour. Ready-mixed mortar or site use of a gauging box at all times will help uniformity. Failure of the pointing or changing mortar colour can lead to expensive remedials. Mortar colour can be changed, not only by changed gauging, but also by the use of a different cement, using wet bricks, or by pointing up the work too soon.

(*h*) Mortar plasticizers can cause problems if incorrectly used. Plasticizers should be used consistently in strict accordance with the manufacturer's instructions. Errors can occur due to:

(*i*) adding plasticizer neat to a batch instead of diluting it first
(*ii*) incorrect dilution of the plasticizer
(*iii*) failure to maintain a constant dilution of the plasticizer.

Problems with brickwork

(*a*) A common problem with external cavity walls is failure due to instability and rain penetration of the wall itself. Instability is caused by using too few wall ties, or if the ties are not long enough to give at least a 50 mm lap on each brick skin, or by pushing the ties into the mortar instead of building them in. To prevent water ingress, ties must fall to the outer leaf, drips be in the centre of the cavity, and no mortar must be allowed to bridge the cavity.

(*b*) Problems can occur due to rising damp, damaging the wall and any floor or wall finishes. To stop this occurring, any moisture from the ground must be stopped from getting inside the building. A damp-proof membrane (DPM) laid under a floor slab must have any joints made with a double welt. The DPM must project sufficiently to overlap the damp-proof course (DPC). Horizontal DPMs must be continuous with vertical steps at changes in direction.

DPCs must be of the specified width, laid on a full mortar bed, with a full mortar bed over them. They should not be placed until the overlying brickwork is placed, thus avoiding any damage due to lack of protection. DPCs should not be pointed over with mortar on exposed faces or bridged by mortar droppings in the cavity. Laps in flexible DPCs should be at least 100 mm.

(*c*) Cavity wall insulation can lead to damp 'bridging' if it is not built into the cavity correctly. Good practice should ensure that the cavities themselves contain nothing which might lead water into the building. It is wrong to build the cavity first and then insert the insulation. One leaf should be built ahead of the other, its inside face cleaned off, the insulation placed and then sealed by the other brick leaf.

Trenches and pipelaying

Trench collapses are a major cause of accidents. Do a careful risk assessment and plan your operations using safe and established practice. The Construction (General Provisions) Regulations 1961 give some guidance on the requirements as follows.

(*a*) An adequate supply of timber or other support materials is to be provided to prevent danger to any person employed from falls of earth, etc.

(*b*) Every part of the excavation should be inspected by a competent person at least once every day when people are working there.

(*c*) No person shall be employed in a trench until an inspection has been carried out by a competent person.

(*d*) Trench timbering shall be carried out by competent workers under competent supervision.

(*e*) Excavations shall be fenced to prevent people falling in.

(*f*) Materials must not be stacked near the edge of a trench so as to endanger those working in the trench.
(*g*) Excavators can be used for lifting providing they have a Certificate of Exemption (CON(LO)/1981/2 General) (HSE, 1981).

Excellent guidance on safe trenching techniques and other considerations is given in the report *Trenching Practice* (CIRIA, 1992). The high degree of danger in trenching and pipelaying operations warrants close attention to all aspects of safety from the onset.

Safety checklist

This is a basic checklist. Other items should be added as appropriate to a particular scheme.

1. Is surface clear of plant, spoil heaps, materials, etc. for at least 1·5 m from the edge of the excavation?
2. Are spoil heaps being properly controlled and will they stay like this in wet weather?
3. Is the trench clear of men while the spoil heap is being worked on?
4. Is the space between the trench and the spoil heap clear of pipes, bricks, stones, tools, etc.?
5. Is the work properly fenced off and 'signed' during the day. Is the work properly fenced off, signed, guarded and lit during the night?
6. Is access adequate without anyone having to jump across the trench. Are footbridges with guardrails available and being used?
7. Are ladders available and being used?
8. Is the supervisor ensuring that no one climbs on the timbering?
9. Is the trench safe from exhaust gases from machines working in the trench or nearby?
10. Does everyone know where the buried services are, and are they clearly marked?
11. Are the men excavating and shoring this trench experienced in this sort of work?
12. Are they working at safe distances from each other?
13. Is the ground as the design assumed?
14. Is there any movement or deterioration of the ground that may put adjacent services, roads or structures at risk?
15. Is the area affected by any blasting or other heavy vibrations?
16. Is the groundwater level as used in the design (i.e. not higher)?

17. Are there proper pumps?
18. Does the pumping arrangement avoid drawing materials from behind the sheeting?
19. Is the work being done in accordance with the drawings or sketches. If not, is the variation permissible?
20. Are unsheeted faces safe, with no sign of peeling away, etc.?
21. Are materials used the correct design sizes and quality?
22. Are wedges tight?
23. Is timbering free of damage by skips?
24. Are waling and strut spacing within ±100 mm?
25. Are deflections excessive?
26. Are all struts horizontal and positioned squarely to the walings (within 1 in 40)?
27. Are frames supported against downward movement (by hangers or lip blocks, puncheons and sole plates)?
28. Have correct pins been used in steel trench struts?
29. Is the method of withdrawing sheeting and support during backfill a safe one?
30. Is work tidy?
31. Are stops provided for mobile plant?
32. Is visibility adequate in trench?
33. Are safety helmets available and being worn?

Trenches need to be as narrow as practicable, but must be adequate to allow the work to progress as quickly as possible. Workmen need room to work effectively in a trench and this can give a trench width requirement in excess of the minimum design requirement. Consideration of the waling and strutting need is required (see Table 4.4).

Table 4.4. Minimum width between walings

Pipe diameter: mm	Trenches not exceeding 3 m deep	Trenches not exceeding 6 m deep
Up to 300	700	1000
300–600	Diameter +400	1000
600–850	Diameter +400	Diameter +600
Over 850	Diameter +600	Diameter +600

The width of wayleave will often be determined by the client. It will vary according to the equipment being used, the requirements for clear access along the line, and the need to store spoil, pipes and other materials clear of the trench. Adequate access is a statutory requirement. Plan the required width and if this can be obtained the job will be made easier as well as safer and quicker. With early planning you may be able to negotiate a widening of a wayleave that you feel to be of a restrictive width.

Many contracts are delayed by incorrect arrangements of the wayleave workspace. The best solution is usually *not* to put the trench down the centreline of the wayleave.

You must ensure that adequate permanent works materials and temporary shoring are available at the commencement of work. This will help speed the works as well as ensure safety. While you need the correct number of workers for each specific trench, you also need adequate plant. This is again a statutory requirement. It also speeds construction. The excavator should *comfortably* excavate to the required invert. If you have a crane in attendance, it should *comfortably* lift the required loads anywhere over the wayleave.

External features to be aware of are underground and overhead services. Expose and protect them ahead of the works. Set up proper 'goalpost' crossings beneath overhead cables.

Barriers around the works must be provided to prevent unauthorised access. The wayleave itself will be fenced and further barriers around the excavation may be required. Make sure stockpiles of pipes cannot be rolled over by children. Deny them access to trenches.

When planning trenchworks in trafficked areas, ensure that excavators and cranes do *not* swing their jibs over trafficked routes unless specific safeguards are in place (banksman, traffic lights). Provide stop boards or traffic barriers wherever traffic can pass close to the trench.

Trench sheets are a common form of side support. They are most effective when pushed into the ground ahead of excavation. They are also 'toed' in below the trench invert. The action of pushing the sheets into virgin ground often puts greater pressure on the sheets than occurs when the trench is excavated and struts and walings are in place. Allow a heavier gauge of sheet to allow the toeing in. Otherwise sheets buckle and distort. The trench is less safe and sheets rapidly become useless.

Consider pumping requirements carefully. Always provide a good sump below the trench formation level. If water seepage is an ongoing problem, provide a carrier drain (100 mm or 150 mm diameter) along the trench bottom and draining into the sump at its lower end. Keep water levels below the trench invert level — your job will then be safer and easier. Do not allow sumps or the pump strainers in them to become blocked with silt. This will clog the pumps. Control of the water ingress will be lost, the trench will flood, work will be stopped. The resultant loss will far outweigh any sump costs.

Ground conditions must be well known before excavation. Daily checks are required during trenching operations to ensure the ground remains what was assumed at the start.

Pipelaying

With rare exceptions, pipes are laid uphill, starting at the lowest point. The pipes are wholly or partially surrounded by fine gravel or concrete depending on the design. When gravel is used, a gravel bed is laid and the pipes 'sat' on this at the appropriate invert level. The remainder of the gravel is laid as pipelaying progresses. With a concrete surround, pipes are packed to level on pre-formed concrete sleepers or bricks, then concreted in.

Pipes with rubber sealing rings are deemed flexible. Flexible joints must be cleaned and greased to allow the rubber sealing ring to work effectively. An approved lubricant should be used. Flexible pipes, laid with a concrete surround, have been designed to give extra strength. It is normal practice to provide flexible joints through the concrete surround. Such joints must be continuous through the surround to ensure flexibility.

Pipelines are usually required to withstand a pressure test. While a full test will occur at periods to suit the works, it is wise to carry out an air test on the line as each pipe is laid. This enables faults in joints to be picked up as they occur. It can be extremely expensive if left until later.

Check each pipe for damage prior to installation. Also check that rubber rings (or gaskets) are not damaged.

Backfilling of trenches

Correct backfilling of the trench is necessary to minimise later settlement.

(*a*) If the excavated material is inadequate it cannot be adequately compacted. In such cases, excavated material to tip and import granular fill.
(*b*) Backfilling in layers of 150 mm thick is generally specified to just above the pipe. 300 mm layers are used above. Ensure good compaction.

Dealing with existing flows (mains)

Many new pipelines follow the line of, and replace, existing lines. Others have junctions with existing lines. When this occurs the added problem of existing flows must be dealt with. In most cases the existing lines are overloaded and this needs to be recognised. The existing flows works must be over-pumped so that the replacement of the old pipes can continue unencumbered. Problems can be encountered and careful planning is required.

(*a*) You must 'tap into' the existing lines above and below the point where you are working. This is best done at times of lowest flow. Monitoring the existing flow is required to establish this time.
(*b*) The seals you provide on the existing pipes, above and below where you are working, must be effective, otherwise you will over-pump and still have to deal with existing flows.
(*c*) It is usual to over-pump from, and to, existing manholes.
(*d*) Over-pumping will be on a 24 hour basis. Electric pumps are effective here as they are quiet and do not require regular fuelling.
(*e*) Over-pumping lines must not leak.

Design considerations include the following.

(*a*) Effective segregation of drainage works from traffic flows. This is necessary for production as well as safety and is an important consideration at the planning stage.
(*b*) Adequate width of wayleaves — in agricultural and similar land. Large diameter pipes and deep trenches demand a wider wayleave than small pipes and narrow trenches. A small increase in wayleave width can often enable much speedier pipelaying to take place. A 20 m width will usually be adequate.
(*c*) Repetitive details. Manholes and the connections and fittings used in drainage work can be standardised to a large degree — pipe and manhole sizes, for example. And the provision of a larger size may be beneficial at a later date.
(*d*) Shallower drainage. Where deep main drainage is involved, it could be beneficial to provide shallow carrier drains on new sections and use backdrops to connect to the deep main sewer. Later work, involving connections to the shallow carrier drains, can be cheaper as a result.

Erection of structures

(*a*) Planning for safe erection should commence at the initial design stage. Essential considerations include:
- (*i*) stability at all stages
- (*ii*) the effect of the erection sequence on stability – where critical, the sequence to be stipulated
- (*iii*) assess loadings at all stages of construction.

(*b*) Provide safe access and working places.

(*c*) Design for ease of component connecting.

(*d*) To ensure safe handling, detail weights and lifting points.

(*e*) Check critical influences independently.

(*f*) The pre-tender health and safety plan should highlight any particular requirements and provide any essential information necessary to allow safe erection.

Chapter 5

The future

Education and training

There is a skills shortage at all levels of the industry and this is posing a threat to the profession's ability to delivery.

Hugh Blackwood (2002), Director of Scott Wilson Railways, writing in *New Civil Engineer*, referred to research on the skills issue carried out by the Railway Industry Training Council:

- the skills shortage is a complex and substantial problem
- increasing pay levels is more a symptom of the growing skills shortage — it is not the cure
- competitive procurement practice leads to duplication and waste as teams compete for contracts by pursuing identical objectives on the same project
- effort is invested in satisfying internal project procedures and financial objectives when the talents of engineers should be harnessed to add value. Today's engineer works harder and produces less, there are quick wins to be had in improving utilisation. Creating meaningful supply chains would help planning and reduce unnecessary competition
- numbers of engineering students under training are diminishing. Colleagues and universities are asked to recruit from a shrinking pool of people; they are given little incentive to offer expensive engineering courses. It is becoming increasingly difficult to find technically included school leavers with adequate mathematics to tackle a course in civil engineering. Education responds by dilution and rationalisation

- the profession has been slow to embrace best practice resource management to ensure that entrants are developed and rewarded as early as possible in their careers
- training is still seen as an undesirable cost in a low margin industry
- access to the profession must be made easier and clearer — paying more is unlikely to solve the problem of entry in the short term.

Blackwood feels that the future is not now dominated by commercial uncertainty. He suggests we need to adopt a growth mentality, to determine what size and mix of skills is needed, and to attract and develop the best people. The industry might consider the following.

(*a*) Graduates enter construction with very good theoretical knowledge but very often little practical experience. What they regard as 'design' is often 'analysis'.

(*b*) Without the relevant practical experience, designs may be progressed which can be inadequate in terms of the construction work itself. This leads to the works progressing more slowly than expected, delayed completions, increased costs to other parties, and perhaps a lower safety achievement.

The Approved Code of Practice (ACOP) to the CDM Regulations 1994 includes a requirement that designers are competent — in paragraph 40(b) the ACOP states that a measure of competence is 'familiarity with construction processes in the circumstances of the project and the impact of design on health and safety'. It could follow that any failure in this basic test of competence is a breach of the regulations.

(*c*) Many contractors experience difficulty in the supervision of specialist sub-contractors. A major reason for that difficulty is that too few engineers understand the construction work involved in the sub-contracts. The relevant drawings often reside unopened in the project office and problems ultimately occur.

(*d*) If the industry is to modernise itself and eliminate the defects highlighted by the Latham (1994) and Egan (1998) reports, fully relevant training and education is necessary. The earlier in an engineer's career that this can be provided the better.

Research and development

(*a*) The events of 11 September 2001, which led to the collapse of the World Trade Centre Towers, has led to careful examination of the use of concrete and steel in such structures. Writing in *New Civil Engineer*, Mike Hitchens (2002), Director at Whitby Bird & Partners, makes reference to the use of concrete:

> Concrete has obvious properties that make it suitable material with which to design tall, robust structures. These are its fire resistance, structural redundancy, monolithic form and stiffness. There are many excellent examples of the use of concrete in recent structures; at Canary Wharf both Harringtons and O'Rourke have built high rise concrete cores using slip form and jump form techniques.
>
> High-rise cores in concrete are inherently stiffer than those in braced steel and can be designed to survive large impact loads and fire. Flat slab construction is consistently providing low cost, short lead in, fast, efficient construction.
>
> Importantly, the simple repetitive flat slab provides a consistent and simple form to which the principle following trades of cladding and building services can be readily fixed with minimal risk of delay or conflict.
>
> At Whitby Bird & Partners we are already building high rise residential and commercial concrete frames based on this simple concept with many more on the way. In Europe concrete buildings are the standard solution with notable major projects built at the rate of two/three floors per week.
>
> To design a robust structure capable of surviving accidental collapse requires more than a suitable choice of construction material. Monolithic in-situ concrete construction goes a long way to ensuring this is achieved almost by default. At the end of the day it is improbable that we can design tall buildings that will survive completely intact.
>
> What we should do is to have a design where key elements are protected from impact or fire — obviously the building core to enable escape and to provide vertical support. The building frame should then be designed such that the loss of an intermediate floor, or floors, does not cause those above to fail. These obvious design requirements can be achieved in concrete or steel by good design for which there is no substitute.

Richard Thieman (2002), Director at Yolles Partnership, referring to the use of structural steel, states:

> The cataclysmic events of the 11 September call for a serious review of structural practice. The disaster showed that a well designed steel building can be remarkably robust under extreme loading and damage. However the final collapse was sudden and

catastrophic. Reports suggest that the collapse was due to failure of both the fire protection and the structural connections between floor trusses and the perimeter columns causing main floor trusses to drop to the floor blow. This reduced the restraint to the main perimeter columns and overloaded the floor below, leading to a domino effect down the building.

Much of the collapse mechanism was independent of the construction material. Columns in any material will fail if they lose the necessary restraint and any floor will fail if sufficiently overloaded. A range of solutions to strengthen high-rise buildings has been proposed such as crash floors and hardened refuge areas and these will apply equally to materials other than steel.

However, fire proofing and connection details were specific to steel. Since the construction of the World Trade Centre fire proofing materials have developed significantly. On current steel high-rise construction in the UK their adhesion to steel has been enhanced by increasing the cementitious content.

In the UK, connection details already need to be designed to resist disproportionate collapse but connections and beams can also be designed to be more ductile and hence better able to absorb the impact from a falling floor.

(*b*) The Transport Research Laboratory is carrying out research work for the Highways Agency on the use of new materials for bridge construction and rehabilitation. Fibre reinforcement polymer (FRP) offers big advantages in terms of its high strength, light weight and durability. Lack of experience and design guidelines has, however, hindered its use in bridges. Research has focused on:
   (*i*) the use of FRP as the main structural element for the bridge itself
   (*ii*) using FRP to strengthen bridge decks and their supports.

(*c*) Pavement material design and production in the UK has tended to follow the lead of other countries. High-tech thin surfacings, used so successfully on UK roads, have their origins mainly in mainland European research and development. Cold mix asphalt is environmentally friendly. It is intended to establish a European centre of excellence in cold mix technology in the North West of England (Nymas Bitumen, Eastham refinery, Liverpool — see Walter (2002) in *New Civil Engineer*).

(*d*) Procurement based on performance-based specifications makes specification itself easier for the design team and allows specialist sub-contractors to prepare details which are based on their specialist knowledge. Study funded by the Department for Trade and Industry examines how relevant data are passed down the supply chain so that informed decision on the choice of materials and components can be made.

(*e*) CIRIA project Report 1844 *Delivering standardisation and pre-assembly for occasional clients* (CIRIA, at press) is intended to show occasional clients and their designers that the benefits of the practice are not confined to large, repeat clients.

(*f*) The BAA/Amec Procurement team have developed a system of precasting to construct service pits beneath taxiways and aircraft standings. The original in-situ product took 300 man-hours to construct on site. The new precast product takes 15 man-hours to construct. Quality is improved, far fewer operative security checks are required and the number of HGV visits to site has been reduced by nearly 50%.

(*g*) Benchmarking. The appropriate software allows you to decide what you want to measure and why, how best to present the data you obtain and how to use them. It can assist businesses in terms of:

(*i*) performance improvement
(*ii*) measurement of best value
(*iii*) supply chain selection and measurement
(*iv*) internal project and divisional comparisons
(*v*) comparison with competitors.

Building Software Ltd, of Tiverton, Devon provide support for many local authorities and companies. They suggest that their product is easy to use, requires little training, and can be customised to suit the client.

Benchmarking Clubs are being established and examples of best practice and other topics can be freely discussed.

From idea to reality

Clients should:

- ensure that they are properly advised as to the appropriate form of contract for a particular project and clarify their needs as far as they reasonably can before they appoint a designer
- consider 'best value' and not just 'lowest price' when carrying out any procurement

- allow adequate time for each part of the construction process — problems tend to occur when clients, for whatever reason, consider their needs to a very late stage and then want to get on with the works as quickly as possible
- ensure all operations are carefully planned — the Japanese take a seemingly inordinate time finalising plans *but* when they start, they expect no hold-ups whatsoever
- ensure that adequate funding is in place and is available at the appropriate time — some projects, the refurbishment of old properties is an example, can be very expensive. A carefully prepared budget is required and it should contain contingency allowances for additional work which may (and is likely to) be revealed
- ensure that their view of the project outcome coincides with that of the designer
- insist on a thorough site investigation and testing process
- consider partnering to construct more cheaply, safely and speedily. Main contractor George and Harding (G&H) design and construct projects for McDonalds. Work is carried out on trust and without a formal written contract. The benefits at the West Bromwich outlet were:
    - capital cost — savings of £125 000 were made through innovative design and construction, £30 000 replicable in future McDonalds
    - construction time — despite the site conditions being more difficult than expected the restaurant actually opened two days early. Value engineering, quick assessments of options and rapid decisions were essential in keeping to budget and programme
    - predictability — McDonalds' projects run efficiently. The supply chain is geared to just-in-time delivery
    - accidents — respect for people is an essential ingredient to make this tight logistical operation a success. There were no accidents at all in the 7000 hours work on this site
    - defects — G&H managers aim for and usually achieve zero defects at handover. Even with the accelerated delivery of the building, they cleared 90% of snags in one day and 100% within four days
    - turnover and profits — G&H has moved onto its fifteenth McDonalds project (at Halesowen, Birmingham), again on a handshake!

Universities in the UK are currently being favoured by an investment boom and carrying out much construction work. Cambridge University takes the view that:

  - ○ tenders are only invited when costs are 80% or more finalised
  - ○ quality is favoured over cost in the proportion 60:40
  - ○ the NEC Engineering Construction Contract (ICE, 1993) is favoured for the spirit of cooperation it engenders
  - ○ the UK lags behind the US in terms of unit cost, buildability, quality and architectural diversity

- ensure that their chosen designers, principal contractors and other contractors appointed have the relevant competencies and resources to execute and complete the project as intended
- the Highways Agency and the Environment Agency are promoting long-term partnering and simplified supply chains. Organisations included in the partnership are assured of a steady income stream from types of work where they are particularly skilled. They are able to keep skilled teams intact and should be able to improve their business as a result. The client is able to employ a known performer who is likely to give a good result. Procurement itself is made easier, and time and effort is saved in the tender process. Early feedback is positive with contractors and designers cooperating from the onset.

Consultants

Designers might:

- be paid a better level of fee to enable them to keep abreast of technology and ensure that their staffs can perform at the top level
- engage on the re-writing of standards so that they cease being like recipe books — allow engineers to engineer and encourage innovation
- ensure that ground investigations are fully adequate — too often this appears not to be the case
- remember that simple solutions are often the best and likely to enable safer and easier construction — complicated solutions will probably be more expensive

- consider who the design is addressing — the man on site who actually carries out the work? He is often the last to be considered
- seek the advice of specialist sub-contractors on how best to incorporate their facilities in the project at the earliest possible date
- consider the inclusion of specialist temporary works as part of the permanent works design, i.e. vertical excavation to be used as the rear shutter of a retaining wall — façade support to be included in the structural steelwork contract
- consider repetitive designs and standardisation of fixtures — bridges, box reservoirs, circular tanks in treatment plants are usually designed as 'one off'. The degree of standardisation in overseas countries tends to be higher than that of the UK. Standardisation is a strong suggestion in the Egan Report (1998)
- give more thought to health and safety, especially in terms of how a task can be carried out safely and easily. This is not a question of regulations, but the possession of practical knowledge of how to carry out work effectively while complying with those regulations. This often leads to a better, often cheaper, product
- prefabricate as much as possible off site. This is likely to improve quality and save on programme time. Extend your thinking to the reinforcement provision where this is practicable
- consider building steel structures 'top down' — building each floor at ground level and jacking it up
- resist pressures to design down to the limit — as well as leaving no margin for error on future modifications, the practice can lead to significant construction problems. Current investigation work on the Thames Barrier could lead to its design life being extended by 70 years. This would be a large saving compared to a new provision
- lightly loaded structures (bridges for car loads only, for example) could benefit by designs which cater for heavier loads — this would allow for future load increases and any temporary access needs
- avoid the use of high maintenance items — painted softwood for example, unless such items are appropriate for other reasons
- consider limited life solutions for short-term problems (use dry bound stone on temporary car parks)
- consider temporary stresses during construction.

Bridges

Practice might consider the following.

(*a*) Increasing the height of bridge decks above the road surface, enabling resurfacing yet still maintaining adequate headroom. Pressure from clients to reduce initial construction costs may preclude this. This can be a false economy. A pleasure barge on its maiden cruise on the newly opened Union Canal in Edinburgh tore off its mast as it passed under an overbridge. The bridge was found to be 50 mm too low.
(*b*) Reduce the number of one-off designs and go for repetition.
(*c*) Where precast beams form the bridge deck:
   (*i*) use precast footpaths and parapets as well — this will increase safety, save several weeks on the bridge construction time, give enhanced quality and reduce overall cost
   (*ii*) the weight and length of the main reinforcement bars which link the precast deck beams can create problems — even a large crane can have difficulty in handling the bars while they are threaded through the beams. Split the bars into two and link them with couplers.
(*d*) Use of permanent formwork where possible, especially on soffits at height. Ensure your design caters for all sensible loads.
(*e*) Provision of access to piers and abutments for maintenance. Without proper access simple tasks can become very difficult. Gantries and rails may be necessary.
(*f*) Provision of properly considered jacking points for bridge bearing replacement needs. Deeper bearing plinths for jacking access.
(*g*) Aluminium parapets are easier to maintain.
(*h*) Cast anchors into the edges and soffits of parapet beams to enable future maintenance access systems to be easily and quickly fixed.
(*i*) Apply silane treatment prior to erection where possible.
(*j*) On railway overbridges construct the parapets integrally with the edge beams and hoist into position as a complete unit.

Road construction

Practice might consider the following.

(*a*) Pinch points occur at junctions. The use of a main flow lane to filter traffic onto/from the main flow increases the problem. Do not reduce the number of main flow lanes. Provide separate filter lanes.

(*b*) Predicted flows on a new road tend to be exceeded and improvements are required to cater for the higher demand. Carriageways are widened, bridges rebuilt and traffic flows re-arranged at considerable expense. The cost of disruption is additional to this. Would an alternative be to build all bridges of adequate span and width to allow future road widening? The cost of extra land and materials to do this would be relatively small and much less than demolition and rebuild later. When road widening takes place, provide further lane(s) alongside the existing carriageway. The benefits will be to:

- (*i*) save demolition and rebuilding of the existing bridges
- (*ii*) minimise disruption to cross-flow traffic
- (*iii*) simplify new lane construction
- (*iv*) minimise delays to main flow traffic
- (*v*) reduce construction time
- (*vi*) save money.

(*c*) Standardise fixtures and fittings.

(*d*) Consider flatter side slopes on cuttings. A 2:1 side slope will require a fence at the top of the cutting. A 15:1 side slope enables the fence to be put next to the carriageway and reduces the land take.

(*e*) Use plastic pipes and gullies. Light and easy to handle, they act as a former to the concrete surround.

(*f*) Always provide positive drainage on carriageways.

Treatment plants

Practice might consider the following.

(*a*) The standardisation of tank diameters (6 m, 9 m, 12 m and so on). This would enable mechanical plant to be made to a standard size and closer tolerances. Standard reinforcement details would ease fixing and inspections. Steel formwork (perhaps client owned) would become advantageous and be more economically used (by repetitive nature of the work).

(*b*) The prefabrication of reinforcement cages for tank hoppers. There would be:
  (*i*) less wasted reinforcement in laps
  (*ii*) reduced congestion in the conical shaping
  (*iii*) improved opportunity for good concrete compaction
  (*iv*) better and more robust steelfixing.

(*c*) The provision of precast settlement tank hopper bases. This would:
  (*i*) give improved quality and a more durable finish
  (*ii*) reduce the weather risk compared to in-situ work
  (*iii*) reduce the excavation by reducing the working space requirement
  (*iv*) eliminate the need for temporary conical formwork.

(*d*) The provision of precast units for settlement tank walls. This would give similar advantages to those listed for settlement tank bases.

(*e*) Standardise fixtures, fittings, pipe and manhole diameters.

(*f*) The provision of modular units to suit plant capacity. This would:
  (*i*) incorporate standard mechanical units
  (*ii*) allow standard details for civil foundations, pipework, bunding and other relevant features
  (*iii*) enable units to be isolated for repair or replacement
  (*iv*) enable transfer to other sites at end of use.

(*g*) Where a supplier is specified, list the relevant product number on the drawings. This would:
  (*i*) ensure correct pricing and ordering
  (*ii*) ease dialogue with suppliers
  (*iii*) assist the feedback process (quote the product number)
  (*iv*) improve familiarity with the product for the future.

(*h*) Standardise the general formwork requirement.

(*i*) Provide bespoke, fully fitted and commissioned pumping situations instead of traditional site construction and assembly.

Concrete box reservoirs

Practice might consider:

- standard reservoir designs — does each have to be unique in itself?
- adequate access — for cleaning out on completion of concreting and later maintenance needs

- standardisation of fixtures, fittings and pipework based on best practice
- precast roofs — to increase quality, speed of construction and reduce cost. Soffit formwork would be eliminated and there would be a durable and inspectable concrete finish.

Drainage

Practice might consider:

- standardisation on pipes, manholes, their sizes and fittings as far as is reasonable
- making provision for future connections when laying new mains — this might enable the avoidance of expensive temporary works and overpumping and make the work safer
- design pipelines in a manner which considers ease of construction — standardise pipe surrounds, for example
- the provision of top reinforcement in manhole slabs
- lay steep drainage runs at shallower gradients and use backdrops at manholes
- when access for maintenance is difficult, use very durable materials
- it is usually preferable to specify the strongest grade of pipe in any given material.

River and marine works

Practice might consider the following.

(*a*) The use of a heavier pile section than the design requires can be advantageous:
  - (*i*) where hard driving occurs, through isolated boulders or cobbles, or where a 'toe in' is needed in hard ground — the heavier section will reduce pile 'crippling' and be more driveable
  - (*ii*) in temporary works where likely future needs are considered

(*iii*) where penetration is not a problem, cantilever piles will eliminate the need for temporary struts and walings — this will save time, save expensive strut and waling provision, improve construction access and simplify the permanent work construction (there will be no obstructions).

(*b*) The use of precast rather than in-situ construction on tidal work — this allows working to the full tidal cycle.

(*c*) Floating plant is expensive to provide and maintain. It may be preferable to construct onshore as much as possible and fix a small number of larger elements. This would be as opposed to piecemeal in-situ work to an extended programme.

(*d*) Avoid winter working as far as possible. It is slower, less safe and more expensive than summer working. Very often contracts are extended into the summer period as a result of earlier winter delays. It could be argued that, as there is a common duty to reduce risk to others, then summer working should be the norm if health and safety regulations are to be complied with.

(*e*) If winter working *cannot* be avoided, due allowance should be made by all parties for the effect this will have on construction work. This tends to be under-estimated.

(*f*) Apply paint finishes to piles and structural steel at works.

(*g*) Use sensible sections and fixings, and work to a sensible tolerance which reflects the conditions being worked in.

(*h*) On gabion work, provide a good stone finish on the exposed face. Quarry waste or a similar product may be adequate as general fill.

General construction

Practice might consider:

- rationalising the number of options considered for each work element — the number of bathroom types, door types and so on
- standardising sizes of elements as much as possible
- speeding construction by using fewer but larger parts — brickwork and blockwork operations required skilled labour and, using small parts, are relatively slow (see *Buildability in Practice* (Ferguson, 1989))
- Bovis Homes are using precast blockwork and considering precast facings to speed house construction — erection is speeded by using large forklift trucks

- simple, repetitive flat slab construction is enabling two or three floors per week to be constructed on some large European projects.

Engineering services

(*a*) There is insufficient knowledge of the engineering services provided in structures at both design and management levels. Mechanical and electrical work presents particular problems. Structural work tends to progress well.

(*b*) It is important to get the basics of a proposed structure right first time and then fit the structure around them. To get the function correctly serviced, and then tailor the appearance to suit this. The reverse is often the case and considerable effort then becomes necessary to improve the functionality. Some clients, retailers for example, know exactly what they want and get it. Other clients, subject to planning and architectural demands, finish with a structure which is acceptable in terms of appearance but whose function has been compromised. Ongoing change becomes necessary to progressively improve the functional need.

(*c*) More training is needed, and this needs to cover an increased number of people. The provision of mentors would help. There is a strong need for more practical input into planning and design processes.

(*d*) More consideration should be given to the layout and design of plant rooms, lifts, staircases and service risers.

Mechanical and electrical installations

Shopping malls

(*a*) In many cases, each shop area is provided by the landlord with either chilled or cooling water flow and return. Each works in conjunction with a ventilation grill exhausting into the central mall area. Experience indicates that neither scenario is adequate for the heat gains which occur when the shops are occupied and trading.

(*b*) Service ducts are often inadequate for the services they have to accommodate.
(*c*) Space for the external equipment made necessary to overcome the shortfall in cooling or ventilation provision by the landlord is at a premium.

Business parks and premises

(*a*) Designs favour appearance rather than function. Space is not provided for external equipment which is regarded as a disfigurement. As most heat gains are from IT equipment and occur throughout the year, in some cases 24 hours per day, the only solution is to open the windows. This is where we started from.
(*b*) In many cases the minimal height of the structural ceiling is inadequate. Most cooling equipment air handlers require 350 mm of space to be available for their installation.

Purchasing

Purchasing, especially when it is carried out on a multi-national basis, tends to be direct from the manufacturer. This by-passes the sub-contractor who is paid solely to install. This adversely affects the sub-contractor in terms of profitability, training is reduced and a future skills shortage becomes inevitable. Smaller businesses find this practice to be widespread and affecting most of the sub-contractors employed. Professionalism and pride in one's work is difficult to maintain and the practical skills available are ignored.

These are harsh words. Yet they state clearly the results of recent procurement policies and give an indication of the likely results of those policies. Such results cannot be of benefit to the industry or anyone connected with it. The basis of long-term partnering is not just to get better results, it is to avoid such scenarios.

Clearly, there is a strong body of opinion which believes that there is scope for improvement of the construction process at all points on the supply chain. If this can be achieved, it will be to the benefit of all those concerned in the process.

References

Barnes, M. (1992). *The CESMM3 handbook*. Thomas Telford, London.

Blackwood, H. (2002). *New Civil Engineer*, 21 March.

Commission of the European Communities Directorate. *Safety and health in the construction sector*. Report by the Commission of the European Communities Directorate — General for Employment Health and Safety Directorate.

Concrete Society and the Institution of Structural Engineers (1986). *Formwork, A guide to good practice*. Concrete Society and the Institution of Structural Engineers.

Construction Industry Council (1994). *Crossing boundaries*. CIC, London.

Construction Industry Training Board. *Construction site safety*. CITB Publications, King's Lynn.

Construction Industry Research and Information Association (1969). *The treatment of concrete construction joints*. CIRIA, London. CIRIA Report 16.

Construction Industry Research and Information Association (1992). *Trenching practice*. CIRIA, London. CIRIA Report 97.

Construction Industry Research and Information Association (at press). *Delivering standardisation and pre-assembly for occasional clients*. CIRIA, London. CIRIA Report 1844.

Egan, Sir J. (1998). *Rethinking construction*. Report of the Construction Task Force on the Scope for Improving the Quality and Efficiency of UK Construction, Department of the Environment, Transport and the Regions, London.

Ferguson, I. (1989). *Buildability in practice.* Mitchell Publishing Co., London.

Health and Safety Executive (1981). *Certificate of Exemption (CON(LO)1981/2 General).* Issued by the HSE under the provision of the Health and Safety at Work Act 1974, the Factories Act 1961 and the Construction (Lifting Operations) Regulations 1961. HSE Books, Sudbury.

Health and Safety Executive (1988). *Health and safety in demolition work. Guidance Notes G729/1–4.* HSE Books, Sudbury.

Health and Safety Executive (1994). *CDM Regulations 1994 and their successor (Approved Code of Practice).* Managing Construction for Health and Safety. HSE Books, Sudbury.

Health and Safety Executive (1995). *Designing for health and safety in construction.* HSE Books, Sudbury.

Hitchens, M. (2002). *New Civil Engineer,* 20 June.

HM Treasury (1997). *No. 3. Appointment of consultants and contractors procurement guidance.* Procurement Group. HMSO, London.

Holroyd, T. M. (2000). *Principles of estimating.* Thomas Telford, London.

Institution of Civil Engineers (1953). *Standard method of civil engineering quantities.* Thomas Telford, London.

Institution of Civil Engineers (1983). *Guidance on the preparation, submission and consideration of tenders for Civil Engineering Contracts.* Thomas Telford, London.

Institution of Civil Engineers (1991). *CESMM3.* Thomas Telford, London.

Institution of Civil Engineers (1993). *The NEC Engineering and Construction Contract.* Thomas Telford, London.

Institution of Civil Engineers (2001). *Membership guidance note. Written assignment for the professional review.* ICE, London.

Joint Contracts Tribunal (1988). *Standard form of building contract. Practice note 20: Deciding on the appropriate form of JCT main contract.* Broadwater Press, Welling Garden City.

Latham, Sir M. (1994). *Construction the team.* Final Report on Joint Review of Procurement and Contractual Arrangements in the UK Construction Industry, HMSO, London.

Maitra, A. (1999). Designers under CDM — a discussion with case studies. *Proceedings of the Institution of Civil Engineers, Civil Engineering,* **132**, No. 2.

Pemberton, A. (2000). CDM Regulations. *New Civil Engineer,* 27 January.

Robens, Lord (1972). *Safety and health at work. Report of the Committee 1970–72.* HMSO, London.

Thieman, R. (20020). *New Civil Engineer*, 20 June.

Walter, J. (2002). Cold Calling. Nymans Bitumen; centre of excellance, Eastham refinery. *New Civil Engineer*, 20 June.

Index